Polymer-Based Functional Nanocomposites
Applications and Prospects

This book offers a thorough insight into polymer-based functional nanocomposites, covering their development, properties, and applications. It describes advanced processing techniques that enhance the mechanical, optical, and photonic performances of these materials. Emphasizing their transformative role, it highlights cutting-edge applications in electronics, water purification, and sustainability. This book also assesses the economic viability and market potential of these materials, bridging the gap between research and real-world impact. Ultimately, it envisions how polymer nanocomposites will drive future innovations and revolutionize materials science.

Key features:

- Provides a comprehensive overview of polymer nanocomposite fundamentals, processing techniques, and property enhancements.
- Highlights advancements in electronics, electrical industries, and sustainability applications.
- Examines the economic viability, industrial potential, and commercialization challenges.
- Explores emerging innovations and the role of nanocomposites in shaping next-generation technologies.

This book provides a comprehensive insight into polymer-based nanocomposites, from fundamentals to industrial applications and future prospects. It is an essential resource for researchers, academics, engineers, industry professionals, innovators, and entrepreneurs.

Polymer-Based Functional Nanocomposites

Applications and Prospects

Edited by

N. B. Singh and Saroj Kr. Shukla

CRC Press
Taylor & Francis Group
Boca Raton London New York

CRC Press is an imprint of the
Taylor & Francis Group, an **informa** business

Designed cover image: Getty Images

First edition published 2026
by CRC Press
2385 NW Executive Center Drive, Suite 320, Boca Raton FL 33431

and by CRC Press
4 Park Square, Milton Park, Abingdon, Oxon, OX14 4RN

CRC Press is an imprint of Taylor & Francis Group, LLC

ISBN: 9781032863689 (hbk)
ISBN: 9781032864587 (pbk)
ISBN: 9781003527602 (ebk)

DOI: 10.1201/9781003527602

Typeset in Times
by codeMantra

Contents

Preface....................vii
About the Editors....................viii
List of Contributors....................ix

Chapter 1 Introduction of Nanocomposites....................1

Saroj Kr. Shukla, Narendra Pratap Singh, and Nakshatra Bahadur Singh

Chapter 2 Polymers and Their Significance in Nanocomposites....................20

Nakshatra Bahadur Singh, Narendra Pratap Singh, and Saroj Kr. Shukla

Chapter 3 Polymer Nanocomposite Processing Techniques....................36

Roli Purwar and Jitendra Kumar

Chapter 4 Enhancement of Mechanical Properties in Polymer Nanocomposites....................51

Deepa Sharma and Purnima Jain

Chapter 5 Optical and Photonic Properties in Polymer Nanocomposites....................78

Deepika Sandil and Vishwa Pratap Singh

Chapter 6 Recent Advances in Polymer Nanocomposites and Their Impact on the Electronic and Electrical Industry....................97

Shashank Sharma and Kalpana Singh

Chapter 7 Functional Polymer Nanocomposites in Water Purification....................116

Ghanshyam S. Chauhan, Sandeep Chauhan, and Kiran Kumar

Chapter 8 Environmental and Green Aspects of Polymer Nanocomposites....................146

Alka Singh and Richa Tomar

Chapter 9 Economic and Commercial Status of Polymer Nanocomposites 166

Mohammad Harun-Ur-Rashid, Abu Bin Imran, and Md. Abu Bin Hasan Susan

Chapter 10 Tomorrow's World: How Polymer-Based Nanocomposites Lead the Nano Revolution 188

Israt Jahan Heme, Kawsar Ahmmed, Sumaiya Hasan Suma, and Md. Abu Bin Hasan Susan

Index 215

Preface

The field of polymer-based functional nanocomposites stands as a cornerstone of contemporary materials science. By merging polymers with nanomaterials, researchers have unlocked extraordinary properties, paving the way for a wide range of applications. This powerful synergy between polymer science and nanotechnology has driven innovation across industries such as healthcare, agriculture, electronics, and energy. *Polymer-Based Functional Nanocomposites: Applications and Prospects* provides an in-depth exploration of this transformative field, encompassing its foundational principles, wide-ranging applications, and future trajectories. Chapters like "Tomorrow's World: How Polymer-Based Nanocomposites Lead the Nano Revolution" delve into the groundbreaking potential of these materials and their role in shaping next-generation technologies. Technical discussions include Chapter 3 "Polymer Nanocomposite Processing Techniques," Chapter 4 "Enhancement of Mechanical Properties in Polymer Nanocomposites," and Chapter 6 "Recent Advances in Polymer Nanocomposites and Their Impact on the Electronic and Electrical Industry," offering a detailed examination of processing techniques, functionalization, characterization methods, and key performance metrics such as mechanical strength, thermal and electrical conductivity, optical behavior, and barrier properties. Sustainability and safety are focal points, with dedicated chapters such as Chapter 8 "Environmental and Green Aspects of Polymer Nanocomposites". Real-world applications are explored in areas such as water purification, textiles, and food packaging, as highlighted in chapters like Chapter 7 "Functional Polymer Nanocomposites in Water Purification." This book concludes with an insightful assessment of the economic and commercial landscape of polymer nanocomposites, offering a holistic perspective on their market potential and long-term impact.

This comprehensive resource is enriched by contributions from leading researchers and experts, ensuring both academic depth and practical relevance. This book is an essential guide for students, researchers, and industry professionals eager to harness the immense potential of polymer-based nanocomposites.

The editors are grateful to all learned contributors.

Prof. N. B. Singh
Prof. Saroj Kr. Shukla
Editors

About the Editors

Prof. N. B. Singh, former Head of the Chemistry Department and former Dean of the Faculty of Science at DDU Gorakhpur University, Gorakhpur, India, is presently an Emeritus Professor in the Department of Chemistry and Biochemistry and is associated with the Research and Development Cell, Sharda University, Greater Noida, UP, India. He is an Alexander von Humboldt Fellow (Germany). His main area of research is materials science. He has published more than 335 research articles and 85 book chapters. He has authored six books and edited eight books. He has been granted four patents. He is the President of the Indian Association of Solid-State Chemists and Allied Scientists, Jammu, India. His name has appeared four times (2021, 2022, 2023, and 2024) in the list of the top 2% of scientists in the world, declared by Stanford University, USA.

Prof. Saroj Kr. Shukla is a Professor in the Department of Polymer Science, Bhaskaracharya College of Applied Sciences, University of Delhi, India. He has published 102 research papers, 25 book chapters, two books, and two special issues. He has received a number of national and international awards for his scientific contributions. His major areas of research include conducting polymers, polymer composites, sensors, and packaging technology.

Contributors

Kawsar Ahmmed
Department of Chemistry
University of Dhaka
Dhaka, Bangladesh

Ghanshyam S. Chauhan
Department of Chemistry
Himachal Pradesh University
Shimla, India

Sandeep Chauhan
Department of Chemistry
Himachal Pradesh University
Shimla, India

Mohammad Harun-Ur-Rashid
Department of Chemistry
International University of Business Agriculture and Technology (IUBAT)
Dhaka, Bangladesh

Israt Jahan Heme
Department of Chemistry
University of Dhaka
Dhaka, Bangladesh

Abu Bin Imran
Department of Chemistry
Bangladesh University of Engineering and Technology (BUET)
Dhaka, Bangladesh

Purnima Jain
Department of Chemistry
Netaji Subhas University of Technology
New Delhi, India

Jitendra Kumar
Department of Polymer Science and Chemical Technology, Department of Applied Chemistry
Delhi Technological University
Shahbad Daulatpur, Delhi, India

Kiran Kumar
Department of Chemistry
Himachal Pradesh University
Shimla, India

Roli Purwar
Department of Polymer Science and Chemical Technology, Department of Applied Chemistry
Delhi Technological University
Shahbad Daulatpur, Delhi, India

Deepika Sandil
Department of Applied Physics
Bhagwan Parshuram Institute of Technology, GGS IP University
Delhi, India

Deepa Sharma
Department of Chemistry
Netaji Subhas University of Technology
New Delhi, India

Shashank Sharma
Department of Chemistry and Biochemistry and RDC
Sharda University
Greater Noida, India

Saroj Kr. Shukla
Department of Polymer Science,
Bhaskaracharya College of Applied Sciences
University of Delhi
Delhi, India

Alka Singh
Department of Chemistry and Biochemistry
Sharda University
Greater Noida, India

Kalpana Singh
Lloyd Institute of Management and Technology
Greater Noida, India

Nakshatra Bahadur Singh
Department of Chemistry and Biochemistry and RDC
Sharda University
Greater Noida, India

Narendra Pratap Singh
Department of Chemistry
U.P. Autonomous College
Varanasi, India

Vishwa Pratap Singh
UNAM – National Nanotechnology Research Center and Institute of Materials Science and Nanotechnology
Bilkent University
Ankara, Turkiye

Sumaiya Hasan Suma
Department of Chemistry
University of Dhaka
Dhaka, Bangladesh

Md. Abu Bin Hasan Susan
Department of Chemistry
University of Dhaka
Dhaka, Bangladesh
Dhaka University Nanotechnology Centre (DUNC)
University of Dhaka
Dhaka, Bangladesh

Richa Tomar
Department of Chemistry and Biochemistry
Sharda University
Greater Noida, India

1 Introduction of Nanocomposites

Saroj Kr. Shukla, Narendra Pratap Singh, and Nakshatra Bahadur Singh

1.1 INTRODUCTION

It is noted that societal developments are linked with the development of new and improved materials. With the continuous advancement of science, materials are increasingly understood in terms of their structure and property. This correlation between structure and property gave rise to a new discipline known as materials science (Singh NB, 2023). The relationship between the processing, structure, properties, and performance of materials is illustrated in Figure 1.1.

Materials in general can be divided into three categories: metals, ceramics, and polymers (Figure 1.2).

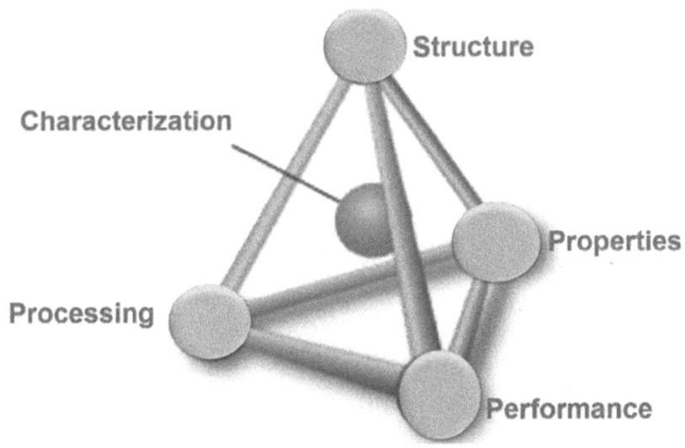

FIGURE 1.1 Processing, structure, properties, and performance of materials.

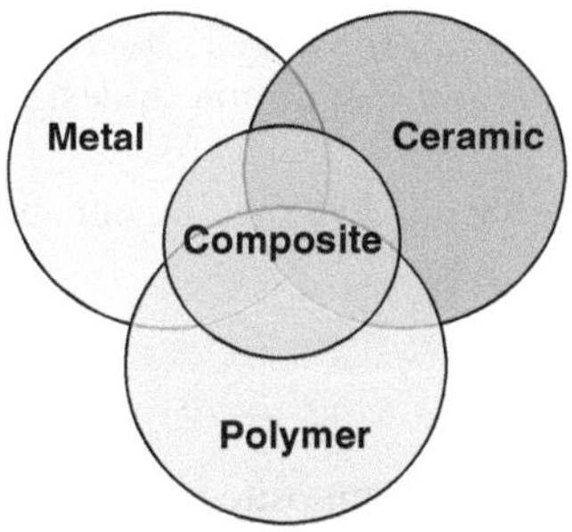

FIGURE 1.2 Categories of materials.

DOI: 10.1201/9781003527602-1

These materials have been used since ancient times for various applications, leveraging their inherent merits and demerits. However, the evolving demands of societal progress necessitate the integration of other properties while also addressing the inherent limitations of traditional materials. An important process is the creation of hybrids or composites from different materials (Khorshidi et al., 2022). The hybrid form of two materials is called as composites, where one serves as a dispersed phase, i.e. matrix, and the other as a filler or reinforcing material. Wood, bone, teeth, molluscan shells, and freshwater pearls are some of the natural composite materials that have served as an inspiration for the development of synthetic composite materials. The synergism between both components produces superior properties compared to the individual components, enabling advanced applications. This also leads to the formation of an interphase region, known as the interface (Silva et al., 2021).

Natural composites have long existed, but they remained largely unnoticed in the past due to technological limitations. For instance, wood is a natural composite with a lignin matrix around cellulosic fibres. Ancient structures using clay mixed with straw or husk are also composites. Natural composites such as oyster shells and spider webs have excellent mechanical strength and stress resistance, often surpassing human-made materials (Mondal et al., 2020). Composites are created by adding reinforcing materials such as fibres, metals, or particles into a matrix, enhancing the material's properties. With growing societal needs and advanced scientific techniques, nanoparticles were introduced. Nanotechnology, which enables the synthesis of metal nanoparticles, is currently very popular. Nanomaterials offer multifunctionality, high performance, and exceptional sustainability, attracting interest across all scientific fields. Advancements in technology and intensified research efforts have led to the development of nanocomposites, positioning them as next-generation engineering materials (Amini et al., 2020).

1.2 DISPERSED PHASE OR MATRIX

The continuous phase of a composite surrounds and binds the dispersed phase and other materials together. Matrix is often a polymer, metal, or ceramic, but mostly different polymers are used as matrix. The careful selection of matrix materials can generate several superior properties. In general, a wide range of composite materials, such as wood and bone, are found in nature, which have inspired scientists and engineers to develop various composites. The basic strategy adopted to integrate natural inspiration into synthetic composites is internal arrangement and structure. Representative examples of natural inspiration in design and material development are listed in Table 1.1.

These natural inspirations have led to specific enhancements in various material properties, as shown in Figure 1.3.

1.3 DISPERSED PHASE

This is the discontinuous phase of composite called as fillers or reinforcing agents. The primary role of the dispersed phase, i.e. reinforced materials, is to optimize or improve the mechanical strength, stiffness, and other desirable properties of the

TABLE 1.1
Representative Examples of Natural Inspiration in Synthetic Materials

Natural Structure	Synthetic Materials	Specific Properties
Fish scale	Kevlar	High tensile strength, heat resistance, low density, durability, and chemical resistance
Lotus leaf	Wind shield	Impact resistance, laminated glass, optical clarity, UV protection, and weather resistance
Bee wax	Building block	Compressive strength, durability, thermal insulation, fire resistance, and acoustic insulation
Reef sharks	Aeroplane	Aerodynamic design, lightweight structure, structural strength, lift and thrust, and environmental tolerance
Birds	Bullet train nose	Aerodynamics, wind resistance, noise reduction, high-speed stability, energy efficiency, and aesthetic design
Insect	Robot	Autonomy, mobility, power source, durability, artificial intelligence (AI), communication, and precision

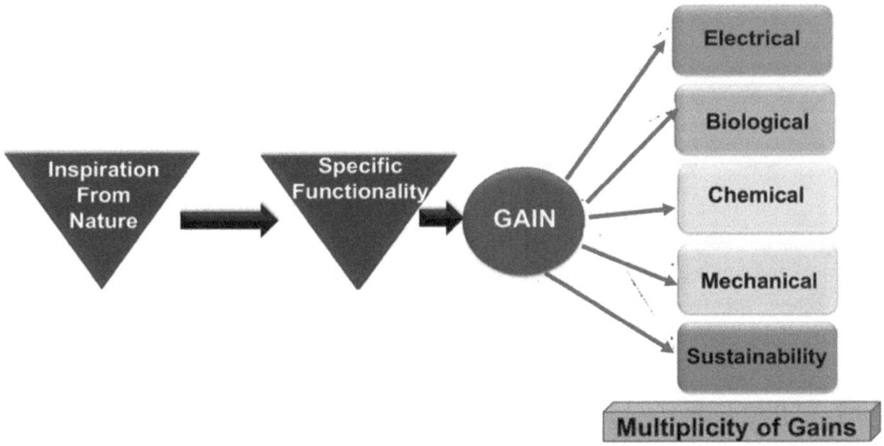

FIGURE 1.3 Impact of natural inspiration on the enhancement of material properties.

matrix for better applications. It can be in the form of fibres, particles, sheets, or flakes of different organic and inorganic materials. The major organic and inorganic fillers are silica, clay, calcium carbonate, carbon particle, and carbon nanostructure. Among the different forms of fillers, fibres are most frequently used because their high aspect ratio between length and diameter enables effective shear transfer between matrix and the fibre, which are made of glass, carbon, or aramid. Examples of fibre polymer composites include fibreglass, concrete, and certain plastics. Currently, dimensional and size optimization in materials adds several advantages in bulk fillers for improvements in various properties of the matrix due to dimensional alignment and larger surface area for better surface interaction and effective interface regions. The composite containing fillers at the nanoscale is called a nanocomposite. Scientifically, the nanocomposite combines two or more materials of which at least one is a nanomaterial (1–100 nm) with different physical and chemical properties. These nanostructured components, often referred to as nanoparticles, nanofibres, nanoplates, or nanotubes, are embedded within a matrix material, which can be a polymer, metal, or ceramic. Nanocomposite materials are designed to exhibit properties that exceed, sometimes drastically, the capabilities of the sum of their constituent parts. The nanoscale phase is within a matrix made of polymer, metal, or ceramic. Adding nanoparticles creates new properties, making nanocomposites superior to individual nanoparticles. The matrix is the major component surrounding the nanomaterial, which acts as a reinforcement or nanofiller. In general, nanocomposites represent a significant advancement in materials science, providing enhanced performance and new functionalities that drive innovation across multiple industries. Nanocomposites are distinct from conventional composites due to their higher surface-to-volume ratio and larger aspect ratio. The nanofiller or reinforcement significantly enhances the material's properties, such as its mechanical properties. For example, the dispersion of clay in nylon significantly improves mechanical strength, making it suitable for use in automobile parts. The addition of clay in polymer also yields the flame retardant nature for better applications due to its oxygen-quenching and charring behaviour (Kumar and Shukla, 2023).

1.4 KEY CHARACTERISTICS OF NANOCOMPOSITES

1.4.1 Nano Reinforcement

The inclusion of nanoscale materials produces a multicomponent structure with novel properties that surpass those of individual components. The interactions and the formation of interphase regions enhance the properties of the matrix for better applications. These reinforcements can provide substantial improvements in mechanical, thermal, electrical, and barrier properties.

1.4.2 High Surface Area

Nanoparticles have a high surface area-to-volume ratio, which leads to enhanced interaction between the matrix and the nanoparticles.

1.5 CLASSIFICATION OF NANOCOMPOSITES

Nanocomposites are categorized based on the matrix and fillers used. The classification of nanocomposite materials can be determined by the matrix, specifically when there is no polymeric material present in the composite. Figure 1.4 illustrates the various types of composites.

The classification of nanocomposites is based on fillers or reinforcing agents used, as shown in Figure 1.5.

The use of different reinforcing agents, i.e. particles, sheets, rods, and fibres, with their inherent features allows the matrix to develop better mechanical, optical, and

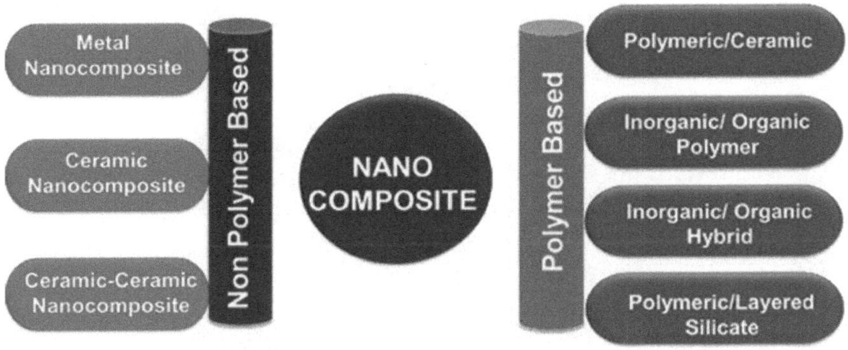

FIGURE 1.4 Classification of nanocomposites.

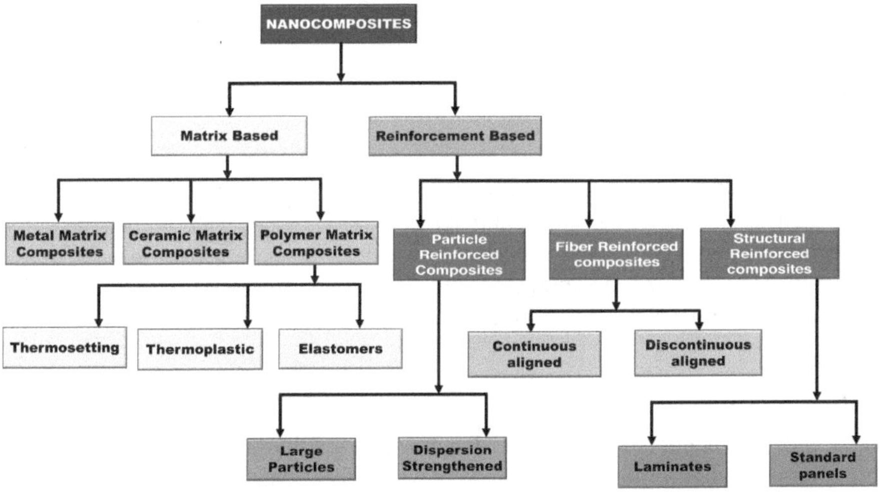

FIGURE 1.5 Classification of nanocomposites based on reinforced agents.

chemical properties. For example, the axial crystallinity of fibres, such as carbon fibre, creates uniaxial mechanical and electrical properties for better applications in fields such as aerospace and automobile engineering.

1.6 PREPARATION OF POLYMER NANOCOMPOSITES

The synthesis of nanocomposites involves combining nanoparticles with a matrix material to create a composite with enhanced properties. The methods used for the preparation of nanocomposites are categorized into direct and indirect methods. In direct methods, both the matrix and fillers are prepared separately and later combined after applying external thermal, electrical, or mechanical energy (Pleşa et al., 2018). However, in the indirect method, the separately prepared fillers are mixed with monomers for chemical functionalization for better polymerization. Further, the functionalized monomers

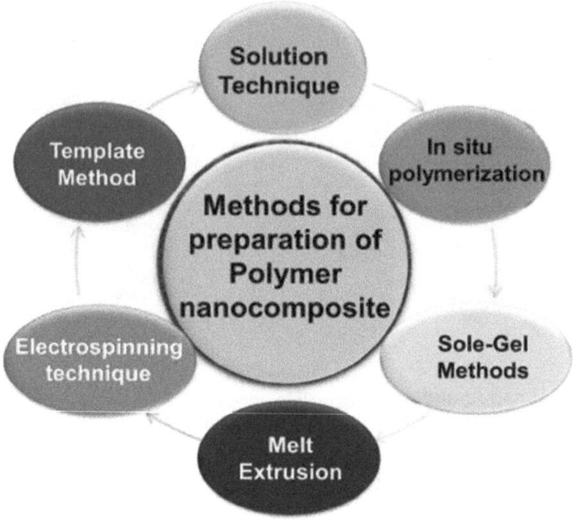

FIGURE 1.6 Methods of preparation of nanocomposites.

are polymerized in the presence of a polymerizing agent, using polymerization techniques such as solution and suspension polymerization. The preparation process can be approached through various methods, each tailored to the specific materials and desired properties of the final nanocomposite (Patil, 2021). An overview of some significant fabrication techniques for nanocomposites is shown in Figure 1.6.

1.6.1 *In Situ* Polymerization

In this method, nanosized fillers are dispersed in a solution of monomer, which is then polymerized to form a polymer matrix with a filler-based heterogeneous matrix. This allows for a uniform distribution of nanoparticles within the polymer matrix, enhancing the embedded properties. Examples include the chemical polymerization of monomers such as styrene, methyl methacrylate, aniline, pyrrole, or vinyl acetate in the presence of nanoparticles such as zinc oxide, silica, clay, or carbon nanotubes (Kumar and Shukla, 2023).

1.6.2 Sol-Gel Process

This is a highly versatile chemical wet approach for the preparation of different types of nanomaterials and structures with better control over morphology and the conversion process. The basic principle of the sol-gel process involves the transition of a solid-liquid solution into solid and gel phases after solvolysis and polycondensation occur. The basic steps involve the dissolution of a metal salt precursor, such as metal alkoxides or metal chlorides, which undergo hydrolysis and condensation reactions to form a network of nanoparticles within the solvent matrix. This method is particularly useful for preparing ceramic nanocomposites with high thermal stability and mechanical strength, especially when more than one precursor salt is used. However, by using different mediums, both organic and inorganic composites can be prepared (Mansour et al., 2022).

1.6.3 Melt Mixing

Melt mixing, or melt compounding, involves mixing nanoparticles with a polymer matrix at high temperatures, above the melting point of the polymer. This method is commonly used for thermoplastic polymers and is carried out using different processing equipment, such as extruders, calendrers, injection moulders, and blow moulders. The high-temperature effect on the matrix ensures good dispersion of nanoparticle fillers within the polymer melt in the absence of solvents, while also minimizing the risk of degradation. Solvent-less mixing preserves the purity of components and maintains chemical stability against solvents. Chemical similarity is an important criterion for melt mixing. For example, polyolefins and paraffines are melt mixed due to their structural similarity. The blending is based on heating of components followed by cooling, which allows structural rearrangement that generates specific features such as optimized glass transition and flexibility of composites.

1.6.4 Solution Mixing

This method is based on solvent-activated interactions between dispersed and dispersion media. In general, solvents functionalize the constituents through a solvolysis process between the solute and solvent. Furthermore, each solute particle interacts with others and yields the composite matrix after precipitation. In solution mixing, both the matrix and the fillers are dissolved in a common single or mixed solvent. After achieving a homogeneous solution, the solvent is evaporated, leaving behind a nanocomposite film or powder. This method offers advantages such as good dispersion and the preparation of the precursor solution for processing into products such as the 3D printer. This method is also effective for producing thin films or coatings for packaging applications. The disadvantages of this method include the use of harmful organic solvents and its time-consuming nature. The use of surfactants and temperature optimization are other features of this method that help prepare nanocomposites with better properties.

1.6.5 Electrospinning

Electrospinning, also referred to as "electrostatic spinning", is a fabrication technique used to produce nanocomposites by applying an external electric voltage to a viscous solution, using the electrohydrodynamic principle. The principle explores electrostatic repulsive forces to produce fibres from a solution or melt after generating an electrostatic jet that overcomes the weak molecular forces of surface tension. The method is classified as horizontal and vertical, based on the position of the melt. The method exhibits advantageous features, including the formation of fibres with scalability, an extremely high surface-to-volume ratio, tunable porosity, and malleability to conform to a wide variety of sizes and shapes. This method has versatile applications for preparing fibres used in filtration, tissue engineering, and sensors. The basic features to consider in electrospinning include concentration, molecular weight, viscosity, surface tension, charge, and conductivity, along with processing parameters, such as applied voltage, distance to the collector, and flow rate (Bhardwaj and Kundu, 2010).

1.6.6 Template Synthesis

This method uses spaces or pores within the matrix of host materials to guide the growth of the required nanocomposite. The spaces lying in the template matrix serve as a pattern to control the morphology and size of the composite structure. The mechanism of morphology control is based on crystal nucleation and the growth of crystal in the available space and dimension. Templates are classified as soft and hard, and they may be composed of natural and synthetic materials. This is a convenient and important method to fabricate the hybrid polymer nanocomposites but is often used with other methods such as sol-gel processing, precipitation, and deposition methods. The advantage of this method is its precise control over the size of nanocomposites. However, the elimination of template after preparation is a tedious process. This method has been widely explored for the preparation of metal oxides, carbon nanostructures, metal nanoparticles, metal organic frameworks, and several polymer composites (Hulteen and Martin, 1997).

1.6.7 Layer-by-Layer Assembly

This technique involves the sequential deposition of several layers of different materials, which can be either organic or inorganic. The basic principle involved in this technique is Langmuir surface adsorption between the deposited layers and Langmuir–Blodgett deposition "LB techniques" introduced three decades ago. This pioneering work typically involves the alternate deposition of positive and negative charged colloidal particles. By controlling the thickness and composition of each layer, highly ordered nanocomposites with tailored properties can be achieved. This method is often used for producing coatings and thin films with specific functionalities (Zhao et al., 2019). The significant factors in layer-by-layer (LBL) techniques are solution concentration, deposition time, the number of layers, and complementary interactions. The formation of heterogenous structures generates complementary surface interactions such as electrostatic forces, hydrogen bonding, and covalent interactions. This layered growth approach with different layers allows for the integration of multiple functionalities, enabling the optimization of interfacial regions and layers for different applications. The notable applications of LBL films are its functional surface coatings, protective coatings, switchable coatings, sensing devices, and energy-related coatings, such as applications used in light-emitting diodes and nanogenerators (Lee and Cho, 2019).

1.6.8 Mechanical Milling

Mechanical milling or ball milling is a solid-state processing technique in which the matrix material and nanoparticles are mechanically ground together in ambient or controlled conditions. In this method, the precursor materials are converted into secondary particles after applying a strong impact to vibrate, rotate, and crush the massive materials into fine particles. The surface impact on secondary fine particles creates flattening, fracture, and rewelding with other particles to produce composite or hybrid materials. The advantageous feature of this method is its low cost and scalability for commercial production. This method can produce nanocomposites with very fine and uniformly distributed nanoparticles, making it a suitable processing tool for materials that are difficult to mix in other ways. For example, mechanical milling is used to disperse different components used for 3D printing and stereolithography of different biopolymer-based nanocomposites. However, certain precautions need to be taken in this method based on the nature of materials, such as brittle and ductile. The temperature also varies in this method, along with the choice of processing medium or matrix material (Yadav et al., 2012).

1.7 CHARACTERIZATION OF NANOCOMPOSITES

Characterization techniques for nanocomposites are critical to understanding their structure, morphology, composition, properties, and performance. These techniques can be broadly categorized into methods for analysing the morphology, chemical composition, structural properties, thermal behaviour, and mechanical properties of nanocomposites.

1.7.1 Morphological Analysis

Morphological analysis techniques provide information about the size, shape, distribution, and interface of nanomaterials within a composite matrix.

Transmission electron microscopy (TEM): TEM provides high-resolution images of the internal structure of nanocomposites, revealing details about the size, shape, and distribution of nanoparticles.

High-resolution TEM (HRTEM): HRTEM can provide atomic-scale information.

Scanning electron microscopy (SEM): SEM gives detailed surface morphology images of nanocomposites, allowing for the analysis of surface features and particle distribution.

SEM combined with energy dispersive X-ray spectroscopy (EDS): This can also provide elemental composition information.

Atomic force microscopy (AFM): AFM provides three-dimensional surface topography at the nanometre scale, which is useful for analysing surface roughness and the distribution of nanoparticles on the surface (Kumar and Shukla, 2023).

1.7.2 Chemical Composition Analysis

These techniques help in identifying the elemental and molecular composition of nanocomposites.

Energy dispersive X-ray spectroscopy (EDS): Often coupled with SEM, EDS helps determine the elemental composition and distribution within the nanocomposite.

X-ray photoelectron spectroscopy (XPS): XPS provides information about the elemental composition and chemical states of the elements on the surface of nanocomposites.

Raman and infrared spectroscopy: These spectroscopy provide information on the molecular structure, phase composition, and interactions within nanocomposites. They are particularly useful for studying carbon nanomaterials and metal oxide-based composites.

Nuclear magnetic resonance (NMR) spectroscopy: NMR provides detailed information on the molecular structure and dynamics of polymers and nanocomposites.

1.7.3 Structural Analysis

Structural analysis techniques are essential for understanding the crystallinity and phase composition of nanocomposites.

X-ray diffraction (XRD): XRD is used to determine the crystalline structure, phase identification, and crystallite size of nanocomposites. It helps in understanding the dispersion and intercalation/exfoliation of layered nanomaterials (Kumar, 2024).

Small-angle X-ray scattering (SAXS): SAXS provides information on the size, shape, and distribution of nanoparticles within a matrix, particularly useful for analysing nanocomposites with polymer matrices.

1.7.4 Thermal Analysis

Thermal analysis techniques assess the thermal stability and properties of nanocomposites.

Thermogravimetric analysis (TGA): TGA measures the changes in weight as a function of temperature, providing information on the thermal stability and decomposition temperatures of nanocomposites.

Differential scanning calorimetry (DSC): DSC measures the heat flow associated with phase transitions, such as melting, crystallization, and glass transitions, giving insights into the thermal behaviour and compatibility of the components.

Thermal mechanical analyser (TMA): TMA determines changes in mechanical properties, such as the glass transition temperature and heat deformation temperature, at elevated temperature.

Dynamic mechanical analysis (DMA): DMA evaluates the mechanical properties as a function of temperature, frequency, and time, providing information on the viscoelastic behaviour of nanocomposites.

1.7.5 Mechanical Analysis

Mechanical characterization techniques are crucial for understanding the strength, stiffness, and durability of nanocomposites.

Tensile and compression testing: These tests measure the strength, modulus, and elongation at break, providing essential data on the mechanical performance of nanocomposites under different loading conditions.

Flexural testing: Flexural tests determine the bending strength and stiffness, which are important for applications where nanocomposites are subjected to bending loads.

Impact strength and surface strength: This test determines the load or impact sustained by a composite material's surface and is useful for structural products such as helmets, bulletproof jackets, and tyres (Kumar et al., 2022).

Hardness: This technique measures the hardness and elastic modulus at the nanoscale, providing insights into the local mechanical properties of nanocomposites. The testing is performed using different tools and standards like Rockwell, Shore, and Durometer.

Abrasion resistance: Abrasion resistance of composite materials is estimated using the standard ASTM protocols after rubbing the surface on a rough surface under a specified load. This test is important to recommend the use of the material in footwear and tyre for scratch resistance applications. The coefficient of friction is an important parameter to compare the different scratch resistance materials.

1.7.6 Other Analysis

Additional techniques can provide more specialized information about nanocomposites.

Electrical conductivity measurements: These measurements are important for nanocomposites intended for electronic applications, providing data on their electrical properties.

Haze and gloss meter: Extent of reflected and absorbed light from the surface of polymers are performed using gloss and haze meter. The measurement was performed using ASTM standards for size and measuring procedure. The properties are used for the comparison of absorbed and reflected light in terms of colour compatibility and light transmission.

Magnetometer: This technique is used to assure the paramagnetic, and diamagnetic nature of polymer composite to explore for different applications.

Bioactivity: The bioactivity of materials indicates the beneficial or adverse effects towards biological reaction and phenomenon, which is important towards the applications for living organisms (Hussain and Thomas, 2021). The basic measuring parameters are biocompatibility, responsiveness, and antimicrobial nature. The basic process and standard are given in Table 1.2.

Measuring bioactivities is a complex phenomenon, which depends on the physical, chemical, biological, and structural parameters (Shen and Fäldt, 2014). The basic measuring principles are based inhibitory effect of materials against biological compounds like enzymes or impact towards the growth of cell or tissues (Lee et al., 2023).

1.8 PROPERTIES OF NANOCOMPOSITES

Nanocomposites are materials that incorporate nanoscale particles into a matrix of bulk material. These nanoparticles can significantly enhance the properties of the bulk material, leading to improved mechanical, thermal, electrical, and barrier properties. The properties of nanocomposites are determined by the nature of the matrix material, the type and distribution of nanoparticles, and the interactions between the

TABLE 1.2
The Basic Process and Standard

S. No.	Test	Standard and Brief Procedure
1.	Biocompatibility	Standard: ISO 10993 provides a framework for biocompatibility testing. **Procedure**: Involves material selection, risk assessment, test selection, sample preparation, in vitro and in vivo testing, followed by analysis and regulatory submission.
2.	Antimicrobial	**Standard**: ISO 22196 for antimicrobial activity on non-porous surfaces. **Procedure**: Involves preparing bacterial cultures, inoculating the material, incubating, and measuring bacterial reduction to evaluate antimicrobial efficacy.
3.	Antioxidant	**Standard**: ISO 15753 specifies methods for determining antioxidant activity. **Procedure**: Involves preparing a DPPH solution, mixing it with the antioxidant sample, incubating, measuring absorbance, and calculating the percentage inhibition of free radicals.
4.	Antibacterial	**Standard**: ISO 22196 for antibacterial activity on non-porous surfaces. **Procedure**: Involves preparing bacterial suspensions, inoculating test materials, incubating, recovering bacteria, and calculating bacterial reduction.

matrix and the nanoparticles (LeBaron et al., 1999). Here are the detailed properties and advantages of nanocomposites.

1.8.1 Mechanical Properties

Strength and stiffness: Enhanced strength: Nanoparticles such as carbon nanotubes (CNTs), nano clays, and graphene can significantly increase the strength of the matrix material by improving load transfer and inhibiting crack propagation. Increased stiffness: The high aspect ratio and excellent mechanical properties of nanoparticles contribute to the stiffness of nanocomposites, making them more rigid compared to the pure matrix.

Toughness: Improved toughness: Nanocomposites often exhibit improved toughness due to mechanisms such as energy dissipation through nanoparticle pull-out, crack deflection, and crack bridging.

Impact strength: Optimum impact strength: The presence of multicomponents in the composite matrix provides the requisite impact strength for the requisite application. The impact strength depends on the type of fillers used and the resulting composite matrix.

1.8.2 Thermal Properties

Thermal stability: Enhanced thermal stability: The addition of nanoparticles can improve the thermal stability of the matrix by acting as a barrier to thermal degradation.

Thermal conductivity: Increased thermal conductivity: Nanoparticles with high thermal conductivity, such as CNTs and graphene, can significantly enhance the thermal conductivity of the composite, making it suitable for applications requiring efficient heat dissipation.

1.8.3 Electrical Properties

Conductivity: Nanocomposites can exhibit enhanced electrical conductivity if conductive nanoparticles such as CNTs, graphene, or metallic nanoparticles are added. This property is useful for applications in electronics, sensors, and electromagnetic interference (EMI) shielding.

Resistance: The resistance of a material is the inverse of conductance and determines the resistance offered by materials in the flow of electrical currents. Impedance is another form of conductance, which indicates the more exhaustive nature of materials to conduct electricity. The resistance of a material can be used to classify it as a semiconductor, conductor, or insulator for appropriate applications.

Capacitance: This property is an indicator for the collection and storage of electrical charge by a material, which is used in energy storage devices such as super capacitor. Currently, this property is also used for sensing applications after measuring the changes in capacitance in response to changes in the surrounding environment.

1.8.4 Barrier Properties

Gas and moisture barrier: Enhanced barrier properties: The inclusion of nanoparticles can reduce the permeability of gases and moisture through the composite material. This improvement is achieved by creating a tortuous path for diffusing molecules, making it difficult for them to pass through the material.

Transparency and opacity: Controlled transparency: Depending on the type and size of nanoparticles, nanocomposites can be designed to be either transparent or opaque. For example, incorporating certain metal oxide nanoparticles can provide UV-blocking properties while maintaining transparency.

1.8.5 Surface Properties

Surface area and reactivity: Increased surface area: Nanoparticles provide a high surface area-to-volume ratio, which can enhance the reactivity and functionality of the nanocomposite's surface. This is beneficial for applications such as catalysis and sensors.

1.8.6 Magnetic Properties

Magnetism: Enhanced magnetic properties: Magnetic nanoparticles such as iron oxide can impart magnetic properties to the nanocomposite, making it useful for applications in data storage, medical imaging, and targeted drug delivery.

1.8.7 Biological Properties

Biocompatibility and antimicrobial activity: Improved biocompatibility: Nanocomposites can be engineered to be biocompatible for biomedical applications, such as tissue engineering and implants.

Antimicrobial properties: Nanoparticles such as silver and copper can provide antimicrobial properties, making nanocomposites suitable for medical devices and packaging materials.

1.9 PROCESSING AND FABRICATION

Processability: Ease of processing: Nanocomposites can be processed using conventional manufacturing techniques, such as extrusion, injection moulding, compression, calendar, and 3D printing. The basic principle for converting nanocomposites into final products is based on the generation of thermomechanical-induced fluidity in the composite followed by solidification and crystallization using shaping units such as mould and dies. Nanoparticles can improve the processability by reducing viscosity and enhancing flow properties due to their smaller size and optimum interparticle interaction, which is used for making a component from nanocomposites.

Scalability: Scalable production: Advances in nanotechnology have made it possible to produce nanocomposites on a large scale, making them more accessible for various industrial applications.

1.10 APPLICATIONS

Nanocomposites, which integrate nanoscale particles into a traditional material matrix, exhibit exceptional properties that make them highly suitable for a broad array of applications (Figure 1.7) (Kiron, 2021)

1.10.1 Automobile and Space

The lightweight nature and optimized mechanical properties of nanocomposites due to better interfacial interactions have been used to fabricate various automobile parts such as hoods, panels, bumpers, doors, mufflers, and engine coatings. The properties are improved by the addition of different organic and inorganic fillers such as nano clays, carbon nanotubes, graphene, activated carbon, metals, and metal oxides. Materials with nanoconfinement significantly improve mechanical properties, enabling better applications in the automobile and space industries. For example, the first use of polymer nanocomposites involved nylon and nano-clay. The basic steps for using polymer nanocomposites are shown in Figure 1.8.

Some of the key properties required for nanocomposites in automobiles include scratch and mar resistance, optimal toughness, improved heat deformation temperature, better dimensional stability, and enhanced impact strength (Garces et al., 2000).

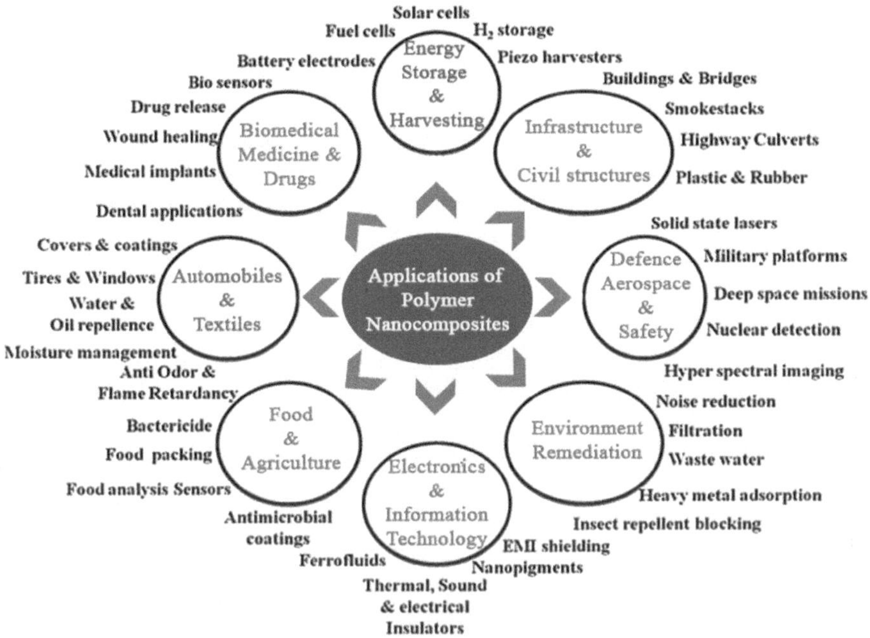

FIGURE 1.7 A schematic diagram listing various applications of polymer nanocomposites.

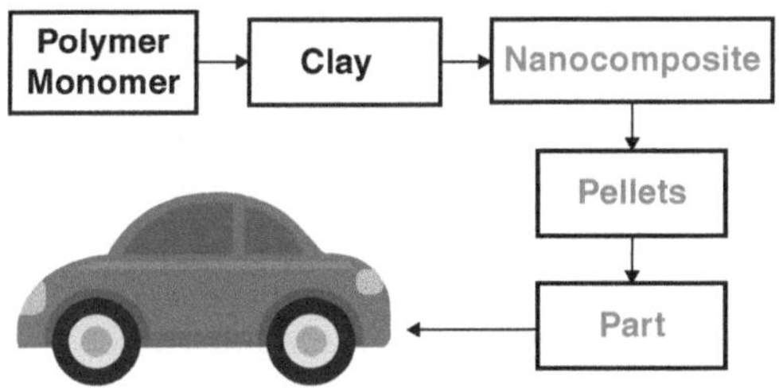

FIGURE 1.8 Key steps to using nanocomposite in automobiles.

1.10.2 Electronics

The optimized electrical and thermal conductivity of nanocomposites makes them suitable for manufacturing various electronic components such as capacitors, resistors, circuit boards, diodes, batteries, smart windows, and memory devices with heat sink properties (Kumar et al., 2024a). The processability of nanocomposites containing polymers, along with their flame retardant properties, make them suitable for making flexible electronic components that support flexible electronic devices for better applications. The development of intrinsic conducting polymers, such as polyaniline, polypyrrole, and polythiophene, makes these composite materials more suitable for electronic devices due to their easily tuneable bandgaps and associated properties. The integration of nanoconfinement in materials channelizes the efficient flow of electrons without energy loss, leading to the development of several advanced materials for various sensing and electronic devices. The important strategy to optimize the band gap and electrical properties is doping of protons, metal ions, and metal oxides. Doping with different dopants disrupts the pi electron clouds around the polymer chain, leading to the formation of polarons and bipolarons, which are responsible for electrical conduction in composites (Dubey et al., 2020).

1.10.3 Agriculture Practices

Polymer composites with biodegradable nature are primarily used in agricultural practices such as seed coating, composition monitoring, soil conditioning, and controlled nutrient and fertilizer delivery, due to their better chemical and physical properties. Biodegradable composites with suitable gas permeability are used for seed coating to preserve the physiological activities of seeds for better yields. Nanoconfinements within these materials also improve degradability and reactivity, making agrochemical applications more effective. For example, blending fertilizers with nanomaterials increases their efficiency and extends their effectiveness over time. The blending of nanotechnology with agrochemicals is referred to as nano fertilizers. The incorporation of biodegradable polymer nanocomposites into soil introduces organic materials and increases porosity, thereby enhancing water retention capacity and improving soil fertility. Furthermore, encapsulating fertilizers within biodegradable polymer coatings allows slow or controlled release for an extended life span. The development of packaging materials is another way to improve the utility of crops and agricultural products. Nature itself provides inspiration, as many natural items are encased in nanocomposite-like packaging, which has motivated scientists to develop nanocomposite-based packaging system for effective and efficient availability of different products. For example, the use of nanosized silica oxide in packaging materials improves their water barrier properties, while the use of nanosized carbon nanostructure improves the mechanical strength, load-bearing capacity, and structural stability of packaging materials.

1.10.4 Biomedical Applications

Medical devices: Nanocomposites with biocompatible and antimicrobial properties are widely used in medical implants, prosthetics, and surgical instruments. The biomedical field is another important area where nanocomposites play a vital role, particularly in drug delivery, biosensing, tissue engineering, and organ implants. The important properties of nanocomposites make them suitable for use in biomedical applications, including biocompatibility, bio-responsiveness, and bio-degradability. Biocompatibility supports the natural growth of biological reactions in the presence of nanocomposite materials (Zheng et al., 2024). This property is integrated by incorporating biopolymers and biologically neutral compounds into the composite. Bio-responsiveness refers to the ability of nanocomposites to change their physical properties in the presence of biological molecules, making them ideal for targeted drug delivery and biosensor applications (Kumar et al., 2023b). For example, grafting chitosan with polypyrrole and polyaniline, in the presence of nanosized zinc oxide, produces a biopolymer nanocomposite that is responsive to hydronium ions. This pH-responsive nature, combined with biocompatibility, makes the composite suitable for sensing of glucose and effective delivery of pH-targeted delivery in cancer treatment. The responsive membranes prepared from nanocomposites are used in potentiometric sensing and fuel cell applications (Rahmani et al., 2022). The multifaceted properties of nanocomposites can also be enhanced by incorporating metallic components such as gadolinium. For instance, a nanocomposite consisting of chitosan and Gd_2O_3 exhibits significant anticancer properties to combat breast cancer and strong antioxidant properties than vitamin C (Khalil et al., 2024).

1.11 CHALLENGES AND CONSIDERATIONS

Despite their numerous advantages, there are challenges associated with the preparation and properties of nanocomposites. The important challenges are listed below:

- *Dispersion of nanoparticles*: Achieving uniform dispersion of nanoparticles in the matrix is crucial for realizing the optimal material properties. The agglomeration of nanoparticles can lead to defects and reduced performance in the final composite.
- *Interface interactions*: Strong interfacial bonding between nanoparticles and the matrix is essential for effective load transfer and property enhancement.
- *Scalability*: The multicomponent and heterogenous structure poses significant challenges for large-scale production. Ensuring consistent quality and performance during upscaling remains a barrier to commercialization.
- *Cost*: The high cost of nanoparticles and the complexity of processing make nanocomposites more expensive than conventional composite materials.

1.12 CONCLUSION

In conclusion, nanocomposites represent a significant advancement in material science, offering enhanced properties that can be tailored for a wide range of applications. Their development continues to evolve, driven by ongoing research and technological advancements across diverse fields such as biomedicine, agriculture, electronics, automobile, and aerospace.

REFERENCES

Amini A, Kazemi S, Safarifard V (2020) Metal-organic framework-based nanocomposites for sensing applications – A review. *Polyhedron* 177:114260

Bhardwaj N, Kundu SC (2010) Electrospinning: A fascinating fiber fabrication technique. *Biotechnology Advances* 28:325–347.

Dubey N, Kushwaha CS, Shukla SK (2020) A review on electrically conducting polymer bionanocomposites for biomedical and other applications. *International Journal of Polymeric Materials and Polymeric Biomaterials* 69:709–727.

Garces JM, Moll DJ, Bicerano J, et al. (2000) Polymeric nanocomposites for automotive applications. *Advanced Materials* 12:1835–1839.

Hulteen JC, Martin CR (1997) A general template-based method for the preparation of nanomaterials. *Journal of Materials Chemistry* 7:1075–1087.

Hussain CM, Thomas S (2021) *Handbook of Polymer and Ceramic Nanotechnology*. Springer.

Khalil KD, Bashal AH, Habeeb T, Abu-Dief AM (2024) Synergistic antibacterial and anticancer activity in gadolinium–chitosan nanocomposite films: A novel approach for biomedical applications. *Applied Organometallic Chemistry* 38:e7531.

Khorshidi M, Asadpour S, Sarmast N, Dinari M (2022) A review of the synthesis methods, properties, and applications of layered double hydroxides/carbon nanocomposites. *Journal of Molecular Liquids* 348:118399

Kiron MI (2021) Kenaf fiber: Properties, cultivation, production, uses and advantages. *Textile Learner* 10.

Kumar A, Kumar S, Tripathi BP (2022) Polydopamine primed phosphorylated sepiolite-polypropylene nanocomposite with enhanced thermal, rheological, and flame retardant properties. *Polymer Degradation and Stability* 202:110005.

Kumar S, Chopra B, Koundal A, et al. (2024a) Scalable chlorination of polypropylene for structural modification and flame retardancy as coating materials of insulation cable. *Polymer-Plastics Technology and Materials* 63: 1861–1873.

Kumar S, Dhawan R, Shukla SK (2023a) Flame retardant polymer nanocomposites: An overview. *Macromolecular Symposia* 407: 2200089.

Kumar S, Kushwaha CS, Singh P, et al. (2023b) Chemiresistive sensor for breath frequency and ammonia concentration in exhaled gas over a PVA/PANI/CC composite film. *Sensors & Diagnostics* 2: 1256–1266.

Kumar S, Shukla SK (2023) Synergistic evolution of flame-retardant hybrid structure of poly vinyl alcohol, starch and kaolin for coating on wooden substrate. *Journal of Polymer Research* 30: 71.

LeBaron PC, Wang Z, Pinnavaia TJ (1999) Polymer-layered silicate nanocomposites: An overview. *Applied Clay Science* 15:11–29.

Lee PH, Lin IH, Hsieh PC (2023) Measurement of bioactive compounds, antioxidants, and α-glucosidase and α-amylase inhibitory effects of Cinnamomum osmophloeum Kanehira extracted with various solvents. *Measurement: Food* 9:100069.

Lee S, Yeom B, Kim Y, Cho J (2019) Layer-by-layer assembly for ultrathin energy-harvesting films: Piezoelectric and triboelectric nanocomposite films. *Nano Energy* 56:1–15.

Mansour NB, Khouqeer G, All NA, Ghoul JE (2022) Synthesis of hybrid nanocomposites by sol-gel method and their characterizations. *Journal of Ovonic Research* 18:57–65

Mondal P, Anweshan A, Purkait MK (2020) Green synthesis and environmental application of iron-based nanomaterials and nanocomposite: A review. *Chemosphere* 259:127509.

Pleşa I, Noţingher PV, Stancu C, et al. (2018) Polyethylene nanocomposites for power cable insulations. *Polymers (Basel)* 11:24

Patil JH (2021) *Nanocomposites – An Overview of Classification and Applications*. JETI.

Rahmani S, Maroufkhani M, Mohammadzadeh-Komuleh S, Khoubi-Arani Z (2022) Polymer nanocomposites for biomedical applications. In: *Fundamentals of Bionanomaterials*. Elsevier, pp 175–215.

Shen JZ, Fäldt J (2014) Requirements of bioactive ceramics for dental implants and scaffolds. In: *Advanced Ceramics for Dentistry*, Shen JZ, Fäldt J (Eds.). Elsevier, pp 279–300.

Singh NB (2023) *Nanocomposites* (Ed. Singh NB). Jenny Stanford.

Silva AO, Cunha RS, Hotza D, Machado RAF (2021) Chitosan as a matrix of nanocomposites: A review on nanostructures, processes, properties, and applications. *Carbohydrate Polymers* 272:118472.

Yadav TP, Yadav RM, Singh DP (2012) Mechanical milling: A top down approach for the synthesis of nanomaterials and nanocomposites. *Nanoscience and Nanotechnology* 2:22–48.

Zhao S, Caruso F, Dähne L, et al. (2019) The future of layer-by-layer assembly: A tribute to ACS nano associate editor Helmuth Mohwald. *ACS Nano* 13:6151–6169.

Zheng Y, Du Y, Chen L, et al. (2024) Recent advances on shape memory polymeric nanocomposites for biomedical applications and beyond. *Biomaterial Science* 12:2033–2040.

2 Polymers and Their Significance in Nanocomposites

Nakshatra Bahadur Singh, Narendra Pratap Singh, and Saroj Kr. Shukla

2.1 INTRODUCTION

The term "polymer" is derived from the ancient Greek words: polus, meaning "many or much", and meros, meaning "parts". It refers to a molecule composed of a molecular structure with several repeating units called monomers. The formation of molecules from a large number of smaller molecules supports the comparatively high molecular mass and related properties compared to other molecules. Thus, polymers are a class of substances with unique properties and large molecular mass, consisting of repeating structural units called monomers, which are connected by covalent bonds through the chemical functionality of monomers (Feldman 2008; Burford 2019). These unique properties have made polymers one of the most adaptable classes of materials, significantly impacting everyday lives over the past several decades by substituting them with other types of materials used in society. However, the increasing demand for materials with optimal properties has compelled researchers to explore possibilities to improve the properties of polymers by incorporating other materials (Shukla 2023; Nepal et al. 2024).

"Happy marriage" between polymers and non-polymers has led to the development of a novel class of hybrid materials with several unique properties for various applications, offering advantages over other traditionally used materials. These hybrid materials are termed as polymer composites (PNCs) or polymer nanocomposites when one of the constituent materials exhibits nanoconfinement in at least one dimension (Nepal et al. 2024). The basic, novel properties of PNCs are the processability of polymers, synergistic mechanical strength, a wide range of electrical properties, induced magnetic properties, and chemical responsiveness. The basic strategies for enhancing the properties of polymers are grafting, composite formation, nanoconfinement, matrix restructuring, and chemical and physical functionalisation. These advancements in polymers have been applied across different industries such as automobiles, textiles, pharmaceuticals, electronics, packaging, and household goods to meet the cost-effective demands of materials for the advancement of society (Khan et al. 2024; Tripathy and Gupta 2024). The findings about these advances in polymers and polymer nanocomposites have been presented by several authors; however, updates are needed for researchers and scientists to design

DOI: 10.1201/9781003527602-2

innovative pathways in this field (Shameem et al. 2021). This chapter is an effort in this direction, which presents the basics of polymers and their importance in nanocomposite materials, based on existing literature. Suitable examples, schemes, and illustrations are included to highlight their indigenous contributions, making the information simple and understandable.

2.2 OVERVIEW ON POLYMERS

Polymers are materials with large molecular mass and are referred to as "macromolecules" due to their large molecular mass in comparison to other materials such as metals, ceramics, pharmaceuticals, and organic and inorganic compounds. Natural polymers such as polysaccharides, proteins, cellulose, and leather have been used by society since the beginning of civilisation in the form of food, clothing, wood, footwear, and packaging materials. However, the controversial macromolecular theory proposed by Hermann Staudinger in 1920 marked a significant shift from the previously accepted agglomeration theory, which explained the nature of high molecular mass substances. Using X-ray analysis of hydrogenated natural rubbers, Staudinger demonstrated the presence of covalent bonds within polymer chains (Frey and Johann 2020). This finding encouraged scientists and industrialists to synthesise a variety of polymers in the laboratory. The proposed structure of natural rubber, as established by Staudinger, is shown in Figure 2.1, along with his photograph.

FIGURE 2.1 Hermann Staudinger with the proposed model of polymers.

TABLE 2.1
Important Discovery in Polymers

S. No.	Scientists and Year	Discovery
1.	Leo Hendrik Baekeland in 1907	Bakelite through a condensation reaction between phenol and formaldehyde, aiming to substitute shellac and hard rubber.
2.	Wallace Hume Carothers in 1930	Nylon fibre was prepared from adipic acid and hexamethylene di amene using the cold drawing method and was called as artificial silk.
3	Eric Fawcett and Reginald Gibson in 1933	Polyethene was prepared accidentally in England after applying high pressure to a mixture of ethylene and benzaldehyde.
4	Paul Hogan and Robert Banks between 1951 and 1953	Polypropylene was prepared through the catalytic polymerization of propylene at low pressure.
5	Rowland Hill and John Crawford in 1930	Polymethyl methacrylate is a transparent polymer and used as a substitute for glass and clear materials. PMMA was also called as Plexiglass.
6	Wallace Hume Carothers and Arnold Collins in 1930	Chloroprene or synthetic rubber was developed by DuPont from dichlorobut-1-ene. It offers several advantageous features over natural rubber.
7	Hideki Shirakawa, Alan Heeger, and Alan MacDiarmid in 1969	Polyacetylene was prepared through the oxidative polymerization of acetylene. It exhibits electrical conductivity and opened up new possibilities for the use of polymers in the electronics industry.

After this breakthrough, several polymers were synthesised in various laboratories by utilising a large number of monomers, polymerization techniques, polymerizing agents, and reactors. Table 2.1 shows important dates and major breakthroughs in polymer research.

The exponential increase in polymer production has led to the development of polymers with a wide range of properties such as tremendous mechanical strength, chemical stability, and durability, enabling them to substitute traditional materials such as wood, fibre, metal, and ceramics. Initially, polymers were used for structural and construction purposes due to their excellent mechanical properties and chemically nonfunctionality. However, by the mid-20th century, environmental aspects and the need for functional and responsive materials prompted both users and manufacturers to explore new capabilities in this novel class of materials. Therefore, since the second half of the 20th century, efforts have been made to enhance polymer functionality through composite formation, polymer chain grafting, and the manipulation of size, dimensions, and structure to enable innovative applications (Ciolino et al. 2022; Hsissou et al. 2021).

PNCs are multicomponent polymeric materials in which at least one of the constituent materials is at the nanoscale, i.e. ranging from 1 to 100 nm. The transition of materials from the macro to the nanoscale imparts enhanced physical and chemical properties, primarily due to increased surface area and a high surface-to-volume

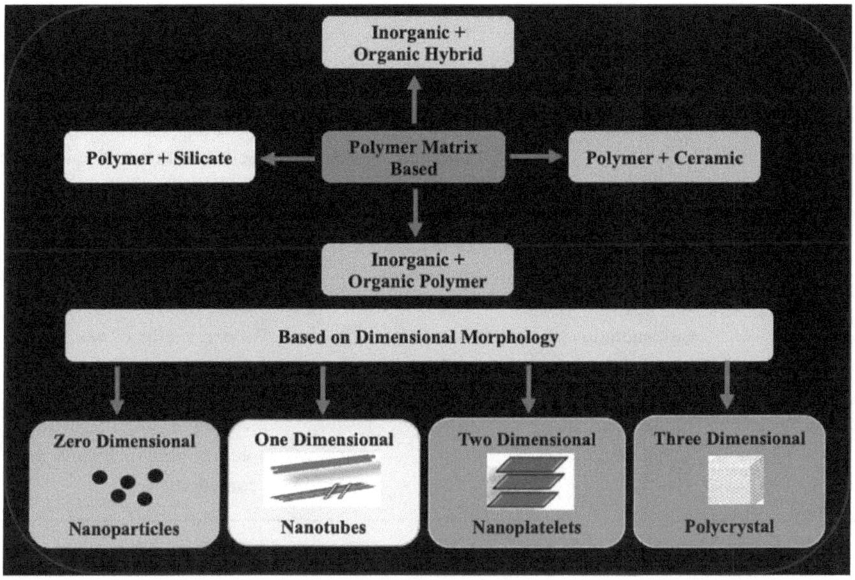

FIGURE 2.2 Classification of polymer nanocomposites.

ratio (Darwish et al. 2022). A brief classification of polymer nanocomposites based on the dispersed phase and dimensions is shown in Figure 2.2.

Although each class of PNCs is used in industry for different applications based on its individual properties, electrically conducting polymer nanocomposites with biocompatibility have been prominently used in biomedical devices such as biosensors, organ implants, and drug delivery systems.

2.3 PREPARATION METHODS

Several methods have been explored for the preparation of PNCs, which are broadly classified as either direct or indirect. In the indirect method, both constituents of the composite are prepared separately and then combined to form the final material. However, in the direct method, monomers are polymerised in the presence of a non-polymeric constituent using a suitable polymerising agent and polymerization processes. The important techniques used for the preparation of PNCs are described in Table 2.2, along with their different features.

Another important point is the processing of PNCs into requisite products. Various methods are used for the processing of PNCs. The basic processing techniques and their brief details are listed in Table 2.3.

2.4 CHARACTERISATION OF PNCs

PNCs are characterised based on their structure, morphology, functionality, and indigenous properties to determine their suitability for specific applications. Properties such as melt flow index, viscosity, and molecular mass are also important

TABLE 2.2
Different Processing Techniques

Techniques	Advantages	Disadvantages	Basic Principle
Solution	Scalable	Impurities	Facile interaction in solution form
Melt compounding	Flexible formulation and economical	Possibility of agglomeration and dispersion	Liquid mixing and recrystallisation
Intercalation	Organised structure	Homogenous dispersion	High-temperature annealing and kneading of components
In situ polymerization	Agglomeration free synthesis and liquid state	Time-consuming and costly	Polymerization of monomer in the presence of non-monomers
Spinning	Good interfacial interaction	Costly	Specialised form of extrusion to prepare fibre in dry and wet conditions

TABLE 2.3
Different Polymer Processing Techniques

Techniques	Brief Details
Injection	This is a technique of mass production to create discrete articles after melting and reshaping plastic using a mould.
Extruder	It is a technique used to convert plastic or metal into a predefined shape.
Compression	Compression moulding is a simple, reliable, and an old processing technique used for preparing sheets through heating and compression.
Rota moulding	Rotating a heated mould containing polymer granules to form a thin shell. It is used to produce parts of great complexity, and with its constantly growing range of applications, rotational moulding is ideal for creating hollow, complex forms like no other process.
Calendaring	This method is used to prepare papers or sheets of polymer by pressing molten polymer between rolls moving at controlled speeds.
Blow moulding	This is a processing technique for the preparation of hollow plastic products such as tubes and bottles.
Thermoforming	It involves the conversion of thermoplastic sheets into the desired shape by softening them through heating and then applying vacuum, pressure, or plugging.
3D printing	The process is based on building up layers of plastic over each other to create the desired shape of a plastic part.

for optimising the processing parameters of polymers and polymer composites (Meyers 2020; Shah 2007). The principles, procedures, and applications of these techniques are available in a large number of books and encyclopaedias. However, a brief overview of commonly used methods for characterising PNCs is given in Figure 2.3.

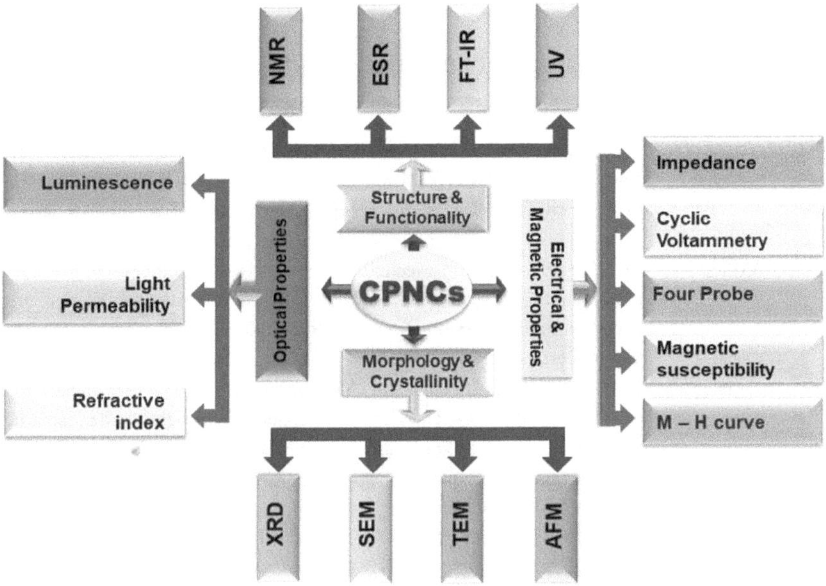

FIGURE 2.3 Schematic representation of the characterisation tool for CPNCs (Kushwaha et al. 2022).

Spectroscopy techniques are in general used to determine the structure and functionality of PNCs. The basic differences among spectroscopy methods lie in the type of radiation used, the energy and interaction patterns, which together decide the inferred material properties. For example, the energy and wavelength of UV-Vis spectroscopy can provide information about the band gap and functional groups present in polymer nanocomposites. The band gap also indicates the probability of electronic transitions in PNCs and provides information about their electrical conductivity and its suitability in electronic devices. The addition of fillers to conducting polymers alter its band gap, thereby extending the application range of conducting polymer-based nanocomposites. Significant information collected from these techniques and their relevance is summarised in Table 2.4.

The physical, chemical, and mechanical properties of PNCs are measured using suitable techniques to confirm their suitability for various applications. The basic properties are shown in Figure 2.4.

The nomenclature of confirming the properties are structural, interfacial interaction, and resultant properties, all of which are important for their recommendation in suitable applications (Verma et al. 2015). A brief discussion of the importance of these different properties of PNCs is provided below.

2.5 THERMAL PROPERTIES

This property indicates the heat-sensitive nature of PNCs in terms of decomposition, structural changes, and heat conductivity. Initially, it was used to confirm the structural stability of materials to ensure their suitability for applications within a specific

TABLE 2.4
Brief Description of the Different Spectrometers Used for PNCs Characterisation

Techniques	Brief Description
UV-vis spectrometer	This technique is based on the optical responses of PNCs to ultraviolet and visible radiation. These responses are used for determining optical barriers, optical sensors, and the presence of functional groups in the constituent.
FT-IR spectrometer	The basic principle of this technique is studying the interaction of infraradiation with matter. This spectrometer has been extensively used for the qualitative and quantitative determination of constituents and their effects on the structure and properties of PNCs.
Raman spectrometer	This is used to identify chemical constituents and evaluate their impact on the structure of PNCs. In general, this technique is considered as a complementary method to infrared spectroscopy for the holistic characterisation of PNCs.
X-ray diffractometer	XRD is a non-destructive laboratory technique used to determine chemical composition, crystal structure, orientation, crystallite size, and layered thickness. The technique is based on the diffraction of X-rays by atoms in the crystal. Common forms of XRD include Powder XRD, Single Crystal XRD, and Small Angle XRD.
Nuclear magnetic resonance spectrometer	NMR spectroscopy is a core technique used for the structural and functional characterisation of polymers and polymer nanocomposites. It works by analysing the magnetic behaviour of nuclei of constituent atoms.
Electron spin resonance spectrometer	This method is used to determine compositional heterogeneity in polymers, along with the generation of free radicals and transitional states, by monitoring the magnetic moments of electrons.
X-ray photoelectron spectroscopy	This is a highly versatile surface analysis technique used for elemental mapping of atoms and molecules present on the surface. The basic principle of XPS involves measuring the binding energy of core-shell electrons present on its surface.

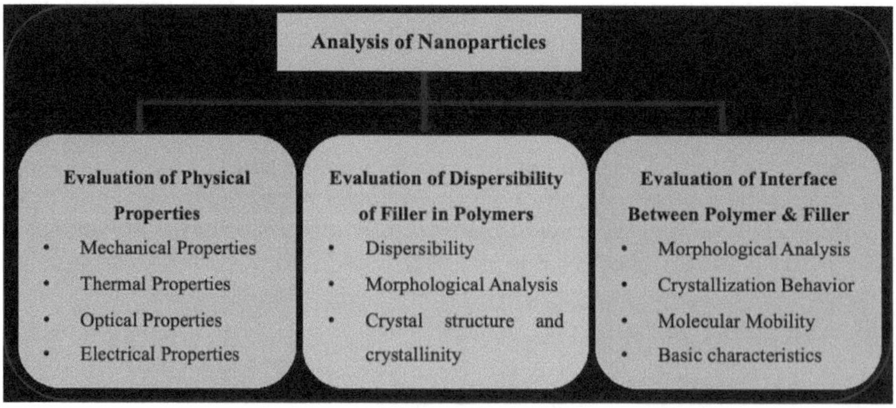

FIGURE 2.4 Different types of analysis for PNCs.

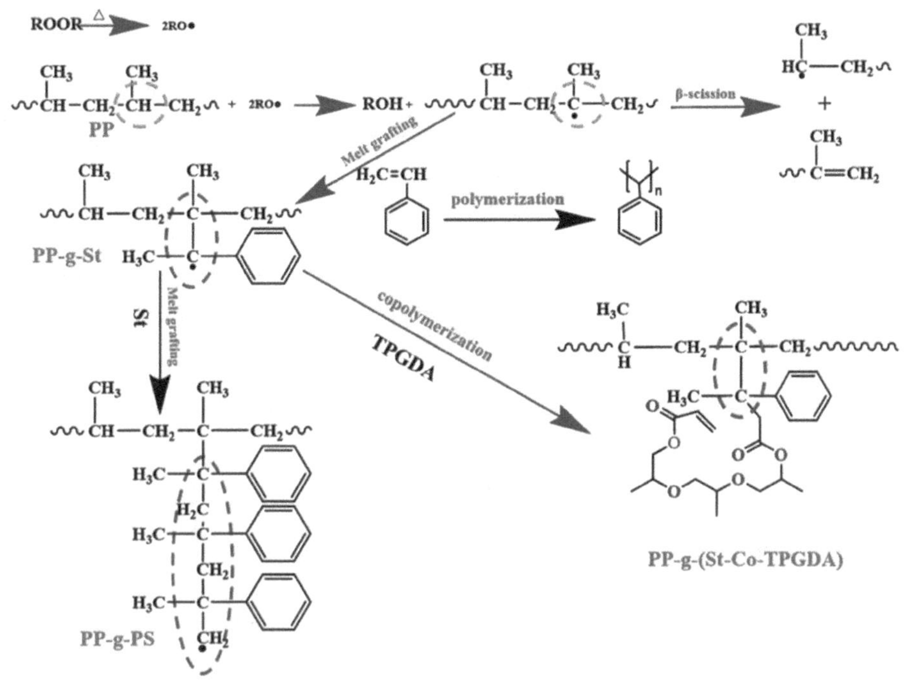

FIGURE 2.5 Scheme for the mechanism of melt grafting of TPGDA and St on PP.

temperature range. Over time, thermal properties have been increasingly used to recommend PNCs for thermal insulating structural parts due to their heat sink and heat-transmitting properties (Yesaswi et al. 2022). Grafing polypropylene with polystyrene and tripropylene glycol diacrylate creates a foam with improved sound and thermal insulation. The prepared styrene-grafted foam exhibits 7.4 times the improved melt strength of pristine PP, along with higher cell density and improved sound and thermal insulation for industrial applications. The underlying grafting mechanism is shown in Figure 2.5 (Liu et al. 2023).

Other significant thermal properties of PNCs are interfacial thermal and mechanical properties, such as melt flow index, Vicat softening temperature, heat distortion temperature, glass transition temperature, and decomposition temperature.

2.6 OPTICAL PROPERTIES

These properties are directly related to the refractive index and extinction index of hybrid polymer medium. Examples of optical properties are absorption, refraction, reflection, and scattering, which are explored for various optical applications such as colour compatibility, gloss, haze, dullness, and turbidity (Srivastava et al. 2008). The basic factors that control the optical properties of PNCs are lattice structure and molecular formula. For example, amorphous structures in general have better optical permeability than crystalline forms. However, the presence of specific functional groups absorbs selective radiation and results in the selective emission of a peculiar group for selective applications. The manipulation of refractive index of polymers is an important development in optical communication, particularly in optical fibres, which exhibit various optical properties. Birefringence, i.e. optical rotation, is another important optical property of PNCs, arising due to anisotropy, i.e. the directional variability of molecular structure. In such materials, the refractive index varies in different directions within the crystal, which rotates the plane of polarised light. The properties of "BIREFRINGENCE" are present in natural crystalline materials but can also be exhibited by synthetic polymers due to the mechanical orientation of the polymer structure.

Scattering of light is another important optical feature of PNCs and related to the optical inhomogeneity of polymeric structures. Here, the molecule acts as a secondary source of light, as proposed by Rayleigh in 1871, and this effect is also referred to as turbidity and total scattered intensity. In polymeric materials, this term is also called "Haze and Transparent" and has several applications in products like clear bags and light barrier advanced packaging materials. The addition of fillers and other additives significantly improves the optical permeability of polymers for various applications. Li et al reported on light management and UV-blocking protocols for diverse photon-related applications, such as energy-efficient buildings, photovoltaics, and optoelectronic devices, using cellulose and lignin composite films (Li et al. 2024). The addition of cellulose nanocrystals in carboxymethyl cellulose exhibits adjustable optical properties such as 90% transparency and haze values between 3.17% and 13%, along with enhanced mechanical strength and thermal stability. Furthermore, the addition of lignin into carboxy methyl cellulose and cellulose films shows excellent UV-blocking capacity by 98.17% for UVB radiation.

2.7 MECHANICAL PROPERTIES

This class of properties refers to the behaviour of PNCs under external static or dynamic stress, leading to changes in dimension and shape. Mechanical properties of polymers reveal strength, brittleness, stiffness, and softness. In general, these are categorised into two types, namely short range and long range, which indicate the tenure of the applied external pressure. These properties are essential to recommend or certify the structural or mechanical applications such as packaging, containers, covers, body parts, and safety devices made from PNCs (Landel and Nielsen 1993). The mechanical properties of PNCs are classified into different categories, which are listed in Table 2.5, along with their brief applications.

The addition of nano-size fillers to the polymer matrix, along with structural manipulation, is another important strategy to optimise various mechanical properties of the polymer matrix (Chandran et al. 2024).

2.8 ELECTRICAL PROPERTIES

The electrical properties of PNCs refer to their dielectric behaviour and electrical conductivity (Warfield and Petree 1961). In general, most pristine polymers are insulators, except for recently discovered conducting polymers such as polyacetylene, polypropylene, polypyrrole, and polythiophene. The basic reason for electrical

TABLE 2.5
Details of Different Mechanical Properties

Properties	Brief Details and Applications
Tensile strength	This is the amount of mechanical stress needed to break a polymer sample with standard dimensions. This property is important for recommending the use of PNCs in applications where they are subjected to stretching, such as in fibres and rubber materials. The basic unit of tensile strength is pascal or $N(M^2)^{-1}$. Polymers with identical tensile strengths may be used in different applications and class depending on their elasticity and stretch behaviour.
Flexural strength	Flexural strength of a polymer indicates the properties of PNCs, which resist deformation under an applied external load. It measures how much stress a material can withstand before bending or breaking. It helps to understand the pressure sustained by 3D structures such as brackets and beams without bending and their end-use applications.
Impact strength	It is the ability of a plastic structure to resist sudden, high-rate loading as seen in applications like helmets and safety gear. The product designed for impact test is checked using different types of impact tests such as Izod, Charpy, dart impact and drop weight tests.
Fatigue	This is a long-term mechanical property that refers to the damage of a polymer caused by repeated cyclic stresses or deformations but its amplitude does not exceed the ultimate strength of the polymer. Fatigue properties are used for checking service failure and structural deformation in polymers.
Creep	This property is also called cold flow, and it indicates the tendency of solid materials to deform slowly under constant persistent pressure. Creep occurs in materials after long-term exposure of high-level stress but still less than the yield stress.
Toughness	This is the property of polymers and hybrid polymers that describes their resistance to deformation and fracture. In general, the area under the stress and strain curve represents the toughness of materials.

conductivity in conducting polymers is the presence of conjugated polymer chains and doping with heterogeneous fillers such as protons, metals, metal oxides, carbon particles, and carbon nanostructures. For example, the doping of ZnO into the polyaniline chain disrupts the conjugated pi electron clouds due to the Lewis acidic nature of Zn. This disruption in pi cloud generates polarons and bipolarons for better electrical conductivity of polyaniline (Blythe and Bloor 2005; Shukla et al. 2012). The difference in band gaps of metal oxides such as ZnO, i.e. 3.5 ev, and PANi, i.e. 1.1 ev, confirms the occurrence of a semiconducting p-n junction for use in electronic devices and sensing applications.

However, pristine polymers exhibit extraordinary electrical insulation properties, with high resistance values ranging between 10^{15} and 10^{18} ohm/cm, making them suitable for applications such as electrical tool handles and electric wire insulation covers. However, the chemical engineering and the dispersion of non-polymeric materials has been explored for the manipulation of electrical properties of polymers. For example, sulphonated polystyrene exhibits significantly improved electrical conductivity in the range of 10^{-6} ohm/cm to 10^{-9} ohm/cm depending on the adsorbed cations. Sulphonated polystyrene is widely used in several industrial applications as a

TABLE 2.6
Significant Electrical Properties

Properties	Brief Details and Applications
Band gap	This is an energetic separation between the filled valance and the empty conduction band. It determines the electrical nature of a material: a small band gap indicates a conductor, while a large band gap indicates an insulator. The method used for estimating the band gap is UV-Vis spectroscopy and the four-probe technique.
Resistance	This property indicates a material's resistance to the flow of electrons. The resistance of a material depends on its structure, temperature, band gap, and activation energy to the flow of electrons. Resistivity meter and probe method are used to measure the resistance of polymers and PNCs.
Conductance	Electrical conductance is the ability of PNCs to conduct electric current in a solid medium. The electric conductance is the inverse of resistance and depends on the band gap and the chemical structure of the polymer. Tuning the conductivity of polymers by forming nanocomposites is an important area of materials to explore the applications of polymers.
Capacitance	The electrical capacitance of PNCs indicates their ability to store electric charge for different applications. The basic unit of capacitance is the farad, including microfarads and nanofarads.
Dielectric strength	This refers to the maximum electric field that PNCs can withstand without experiencing dielectric breakdown or becoming electrically conductive. The variable dielectric strength of PNCs has been explored for a wide range of applications in energy storage and electrical separation devices (Pandey, J. C., & Singh, M. (2021). Dielectric polymer nanocomposites: past advances and future prospects in electrical insulation perspective. *SPE polymers*, 2(4), 236–256).

semiconducting polymer for sensors and membranes (Wallace 1973). The important electrical properties of polymers are band gap, resistance, capacitance, and conductance. A summary of these properties is listed in Table 2.6.

2.9 POTENTIAL AND PROSPECTS

The associated features and properties of polymers and polymer nanocomposites have been explored for different applications.

2.9.1 Aerospace

The relatively inexpensive nature and tunable properties of PNCs have been widely explored in the aerospace industry for manufacturing structural components and electronic devices. The important property of PNCs in aerospace applications is their multi-functionality, which is developed after blending with different nanofillers. The important properties of PNCs are given as follows:

a. Optimal nano-sized confinement within the polymer matrix
b. A wide range of enhanced properties, i.e. mechanical, electrical, optical, magnetic, biological, and electronic, due to the dispersion of nanofillers

c. The formation of larger interfacial areas due to the arrangement of nanoparticles

For example, CNT-based composites have been used in the fabrication of structural parts, such as the rotating blades of aeroplane turbines, due to their excellent rigidity and ability to maintain the cross-sectional profile of the blade. The basic structure of the proposed blade is shown in Figure 2.6 (Njuguna et al. 2012).

2.9.2 Automobile

An important user of PNCs is the automobile industry, where they are employed in the manufacturing of various components. Different parts of a car body made from PNCs, prepared from various fabrication routes, are shown in Figure 2.7 (Shah et al. 2022).

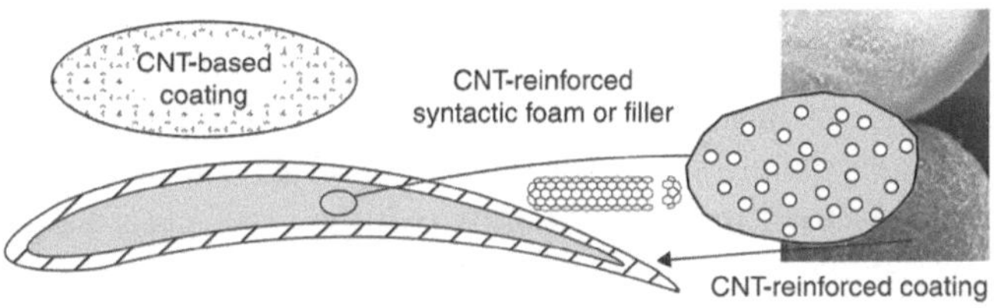

FIGURE 2.6 CNT and polymer composite materials for aircraft engine blades.

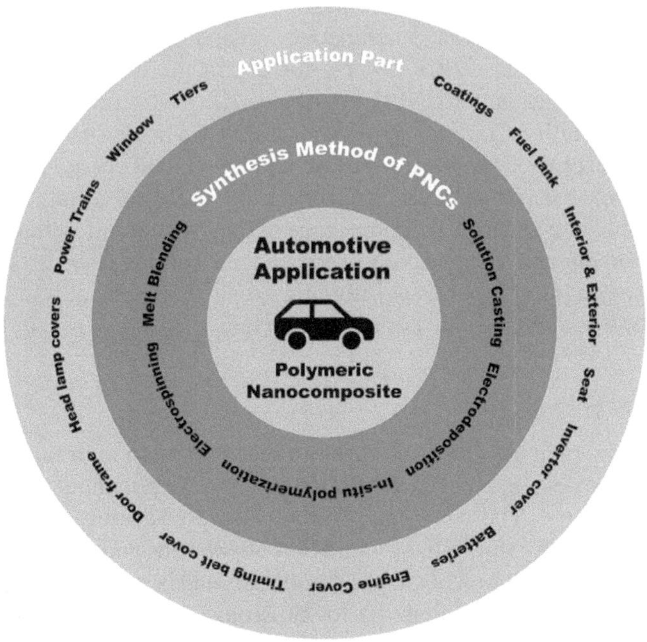

FIGURE 2.7 The schematic representation about the use of PNCs in automobiles.

TABLE 2.7
Representative Polymer Nanocomposites and Their Applications in the Automobile Industry

Polymers	Brief Details and Applications
Polyamide	Polyamides are utilised in a variety of applications in the automobile sector, including gasoline tanks, reservoirs, tubing, pipes, carburetors, injectors, and manifolds.
Polyolefines	Polyolefines are used in various exterior vehicle applications, including door frames, seat backs, hood louvers, heavy-duty electrical enclosures, sail panels, fascias, box rails, middle bridges, step-assists, rocker covers, vertical and horizontal body and closure panels, side trims, instrument panels, grilles, and fenders
Polyurethane	Weather-resistant automotive coating
Rubber	Tire applications
Epoxy	Anti-wear, anti-corrosive, and barrier coating

The advantageous features of PNCs are lightweight nature, excellent thermal and mechanical strength, corrosion resistance, and ecofriendliness (Shah et al. 2022). Several polymers are used in the automobile industry, and representative examples, along with their specific uses, are listed in Table 2.7.

2.9.3 Construction

The innovative properties of polymer nanocomposites remain a relatively under-explored area, despite their vast potential for research and commercialisation. The basic properties that make PNCs suitable for use in the construction industry are insulation towards heat and sound, long service life, anti-corrosive nature, strong adhesion, and durability. Common applications of PNCs in construction include their use in concrete, asphalt, adhesives, coatings, insulation, and energy (Feldman 2014). PNCs used in construction are heterogeneous, multiphase solid state materials reinforced with nano-size agents. The importance of these nanoreinforcing agents lies in their fine dispersion and the ability to fine tune their surface chemistry to exchange the properties. The important nanoreinforcing agents and their roles in the construction industry are summarised in Table 2.8.

2.9.4 Electronics Devices

The use of polymers in electronic devices is a recent expansion in this field, particularly following the development of electrically conductive polymers, which are called "conducting polymers". These polymers contain disrupted conjugate double bonds for conduction of electricity through the formation of polarons and bi-polarons. The dispersion of metal oxides into a conducting polymer matrix introduces heterogeneous properties that enhance their suitability for electronic devices. Shukla et al. reported the synthesis of a tin oxide and polyaniline nanocomposite with improved electrical conductivity and a metallic centre. These synergistic properties are explored for

TABLE 2.8
Representative Fillers and Their Use in Construction Industries

Reinforcing Agent	Brief Details and Applications
Clay	It has been used to improve the mechanical strength of various polymers such as epoxy, polyamide, and polypropylene for different applications.
Silica	Silica, including, nanosilica and its colloid forms, is used as a filler in the polymer industry. It has important applications in concrete industries, where it improves properties such as density, porosity, and workability.
Carbon nanostructure	Graphene, graphite, and carbon nanotubes are key examples of carbon-based nanostructure fillers. They differ primarily in their dimensional structure and frequently used for improving mechanical strength, surface properties, and chemical stability. These fillers also contribute to flame retardancy, making them suitable for structural applications.
Metal oxide	Titanium oxide and zinc oxide are the most widely used fillers in the polymer and rubber industries. Apart from improving the mechanical properties, these oxides are also used as pigments and stabilisers to improve the structural stability of the materials. ZnO also acts as an activator and plays a crucial role in the vulcanisation process of rubber in the rubber industries.

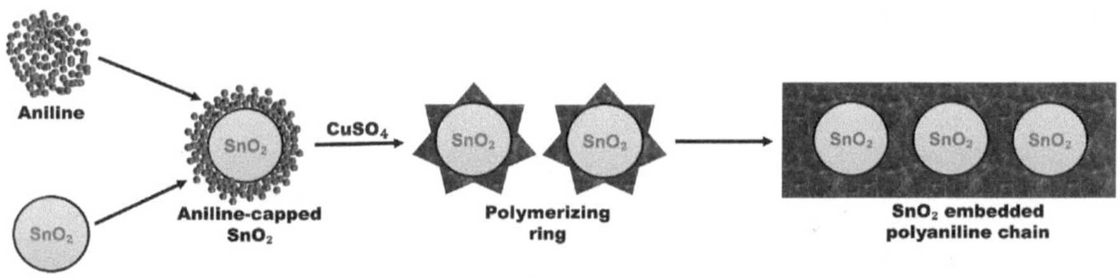

FIGURE 2.8 Preparation scheme for SnO_2 and polyaniline nanocomposites.

humidity sensing across a broad range of 5–95% RH with effective sensing parameters, i.e. a sensitivity of 0.22% RH^{-1}, a response time of 26 s, and a recovery time of 30 s. The scheme for preparing the composite is shown in Figure 2.8, along with its technical details (Shukla et al. 2016).

The other representative applications of PNCs for the preparation of electronic devices are given in Table 2.9 (Tadesse et al. 2024).

2.10 CONCLUSION AND PROSPECTS

The fundamental properties of polymers are described along with their importance in the development of various polymer nanocomposites. The basic properties of PNCs are also discussed, along with their testing procedures, including spectroscopy, microscopy, and methods for evaluating electrical and optical properties. Finally, brief applications of PNCs in aerospace, automobile, construction, and fabrication of devices are discussed, along with their features and limitations.

TABLE 2.9
Representative Polymer NanoComposites for Electronic Devices

Name of Device	Brief Details and Applications
Solar cells	The innovation of tunable bandgaps has been explored for preparing flexible and organic solar cells with several advantages.
Light-emitting diodes	Several fillers, such as perovskite-doped organic polymers, have been used in the development of different types of light-emitting diodes.
Capacitor	Structural and compositional optimisation of polymers such as polyaniline and polythiophene has been used in the preparation of capacitors and other energy storage devices. The optimised properties of polymer for use in capacitors include conductivity, charge carrier mobility, band gap tuning, and pi electron delocalisation.
Nanogenerator	Compressibility and surface-interacting conductivity of polymer matrices have been explored for the development of piezoelectric nanogenerators, which can be integrated into devices for self-powered applications.
Wires and cables	Electrical conducting polymers such as polypyrrole have been used in low-voltage power supply systems, particularly in aircraft.

REFERENCES

Blythe, A. R., & Bloor, D. (2005). *Electrical Properties of Polymers*. Cambridge University Press.

Burford, R. (2019). Polymers: a historical perspective. *Journal & Proceedings of the Royal Society of New South Wales, 152*(2):242–250.Chandran, A. J., Rangappa, S. M., Suyambulingam, I., & Siengchin, S. (2024). Micro/nano fillers for value-added polymer composites: A comprehensive review. *Journal of Vinyl and Additive Technology, 30*:1083–1123. https://doi.org/10.1002/vnl.22106.

Ciolino, A. E., Gonzaga Oliveira de Freitas, A., Satti, A. J., & Ninago, M. D. (2022). From atoms to macromolecules: 100 years of polymer Research. *Advances in Organic Synthesis, 16*:212.

Darwish, M. S., Mostafa, M. H., & Al-Harbi, L. M. (2022). Polymeric nanocomposites for environmental and industrial applications. *International Journal of Molecular Sciences, 23*(3):1023.

Feldman, D. (2008). Polymer history. *Designed Monomers and Polymers, 11*:1–15.

Feldman, D. (2014). Polymer nanocomposites in building, construction. *Journal of Macromolecular Science, Part A, 51*(3):203–209.

Frey, H., & Johann, T. (2020). Celebrating 100 years of "polymer science": Hermann Staudinger's 1920 manifesto. *Polymer Chemistry, 11*(1):8–14.

Hsissou, R., Seghiri, R., Benzekri, Z., Hilali, M., Rafik, M., & Elharfi, A. (2021). Polymer composite materials: A comprehensive review. *Composite Structures, 262*:113640.

Khan, S. B., Chen, S., & Sun, X. (2024). Advancements in polymer nanocomposite manufacturing: Revolutionizing medical breakthroughs via additive manufacturing. *Polym. Bull. 81*:1–53.

Kushwaha, C. S., Singh, P., Shukla, S.K., & Chehimi, M. M. (2022). Advances in conducting polymer nanocomposite based chemical sensors: An overview. *Materials Science and Engineering B, 284*:15856.

Landel, R. F., & Nielsen, L. E. (1993). *Mechanical Properties of Polymers and Composites*. CRC Press.

Li, X., Li, J., Shen, X., Cao, M., Wang, Y., Zhang, W., & Xu, F. (2024). Transparent cellulose/lignin composite films with adjustable haze and UV-blocking performance for light management. *ACS Sustainable Chemistry & Engineering*, *12*(14):5427–5435.

Liu, F., Shen, C., You, F., Zhao, W., Deng, C., & Jiang, X. (2023). Enhancing the sound and thermal insulation properties of polypropylene foam by preparing high melt strength polypropylene. *Macromolecular Rapid Communications*, *44*(20):2300344.

Meyers, R. A. (Ed) (2000). *Encyclopedia of Analytical Chemistry: Applications, Theory, and Instrumentation*. Wiley and Sons.

Nepal, D., Haines, J., & Vaia, R. A. (2024). Polymer nanocomposites: 35 years on. *MRS Bulletin*, *49*(3):236–246.

Njuguna, J., Pielichowski, K., & Fan, J. (2012). Polymer nanocomposites for aerospace applications. In: *Advances in Polymer Nanocomposites*, Fengge Gao (Ed.). (pp. 472–539). Woodhead Publishing.

Shah, V. (2007). *Handbook of Plastics Testing and Failure Analysis*. Wiley and Sons.

Shah, V., Bhaliya, J., Patel, G. M., & Deshmukh, K. (2022). Advances in polymeric nanocomposites for automotive applications: A review. *Polymers for Advanced Technologies*, *33*(10):3023–3048.

Shameem, M. M., Sasikanth, S. M., Annamalai, R., & Raman, R. G. (2021). A brief review on polymer nanocomposites and its applications. *Materials Today: Proceedings*, *45:*2536–2539.

Shukla, S. K. (2023). Century impact of macromolecules for advances of sensing sciences. *Chemistry Africa*, *6*:5–37.

Shukla, S. K., Shukla, S. K., Penny, P. G., & Agorku, E. S. (2016). A resistive type humidity sensor based on crystalline tin oxide nanoparticles encapsulated in polyaniline matrix. *Microchimica Acta*, *183*:573–580.

Shukla, S. K., Vamakshi, M., Bharadavaja, A., Shekhar, A., & Ashutosh, T. (2012). Fabrication of electro-chemical humidity sensor based on zinc oxide/polyaniline nano composite. *Advanced Materials Letters*, 3(5):421–425.

Srivastava, S., Haridas, M., & Basu, J. K. (2008). Optical properties of polymer nanocomposites. *Bulletin of Materials Science*, *31*:213–217.

Tadesse, M. G., Ahmmed, A. S., & Lübben, J. F. (2024). Review on conductive polymer composites for supercapacitor applications. *Journal of Composites Science*, *8*(2):53.

Tripathy, D. B., & Gupta, A. (2024). Nanocomposites as sustainable smart materials: A review. *Journal of Reinforced Plastics and Composites*, 07316844241233162.

Verma Balani, K., Agarwal, A. & Narayan, R. (2015). Physical, thermal, and mechanical properties of polymers. *Biosurfaces: A Materials Science and Engineering Perspective*, *329*, Wiley.

Wallace, R. A. (1973). Electrical conduction in sulfonated polystyrene films. *J. Appl. Polym. Sci. 17*(1):231–238.

Warfield, R. W., & Petree, M. C. (1961). Electrical resistivity of polymers. *Polym. Eng. Sci. 1*(2):80–85.

Yesaswi, C. S., Satya, S. K., Sahu, S. K., Badgayan, N. D., Murthy, P. S. R., Kumar, V. R., & Sreekanth, P. R. (2022). Thermal properties of polymer nanocomposites. In: *Advanced Polymer Nanocomposites* (pp. 99–143). Woodhead Publishing.

3 Polymer Nanocomposite Processing Techniques

Roli Purwar and Jitendra Kumar

3.1 INTRODUCTION

Traditionally, additives are used in the polymer matrix to enhance its processability and functionality. Inorganic particulate additives are widely used in polymers to enhance their functional properties and reduce the cost. Over the past four decades, it has been observed that reducing the dimensions of these particulate additives to nanosize can result in exceptional surface and bulk properties. Such composites are called hybrid inorganic–organic nanocomposites. The incorporation of nanoparticles such as carbon nanotubes, graphene, POSS, nanoclays, metals, and metal oxide has improved the structural, thermal, mechanical, and functional property of polymers for their use in industrial sectors and has paved the way for continuous growth in polymer markets (Zahran and Marei 2019; Potts et al. 2011; Zhanga et al.2017; Peponi et al. 2014; Zhua et al. 2019; Joshi et al. 2018). The theoretically excellent properties achieved by adding these nanoparticles can only be realized when they are completely dispersed in the polymer matrix in their nanodimensional form, creating a large interfacial surface area. A physical mixture of polymer and nanofiller does not lead to the formation of a nanocomposite. During mixing, the nanofiller particles agglomerate due to their cohesive nature and high surface energy, resulting in the formation of microparticles within the polymer matrix. In such cases, fillers and polymers separate into discrete phase and behave like microcomposites. For example, when nanoclay (layered silicate) is mixed with a polymer, the degree of polymer penetration into the clay galleries results in two different types of morphologies: intercalated and exfoliated (Figure 3.1). In the intercalated structure, polymers and clay are present in an ordered form by only a few nanometres. The clay layers exhibit an expansion of less than 20–30 Armstrong due to the penetration of polymer chains. In exfoliated structure, clay platelets (tactoids) are delaminated due to the penetration of polymer chains and are completely dispersed within the polymer matrix. The properties of exfoliated nanocomposites are superior compared to those of intercalated nanocomposites (Zhanga et al. 2017). Different strategies need to be followed depending on the type of polymer and inorganic nanoparticles. This chapter focuses on the various approaches required to process inorganic nanoparticles into thermoplastic and thermosetting polymer matrices. The processing of polymer nanocomposites can be divided into three major steps: exfoliation of the nanofiller, compounding of the nanofiller in the polymer matrix, and fabrication of articles. The nanocomposite articles are

 DOI: 10.1201/9781003527602-3

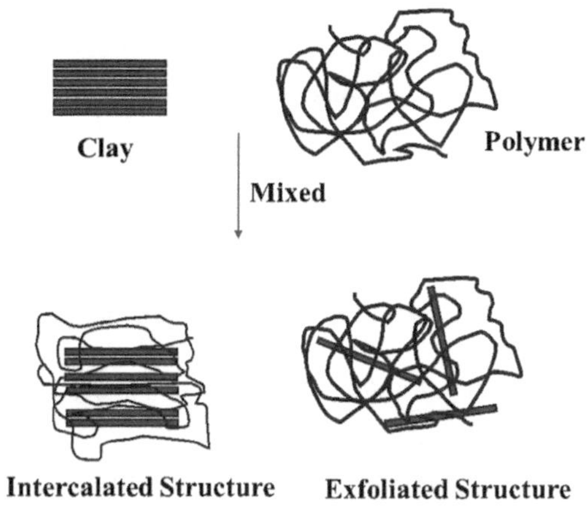

FIGURE 3.1 Schematic diagram for intercalated and exfoliated morphology of polymer/clay nanocomposites.

produced in the form of films, fibres, nanofibres, hydrogels, as well as 3D-printed and moulded products using conventional polymer processing techniques. The first two steps are discussed in detail in this chapter. Fabrication techniques such as 3D printing and electrospinning for the formation of nanocomposites are also discussed in this chapter.

3.2 EXFOLIATION OF NANOFILLERS

Exfoliation of nanofillers can be achieved through three different methods: (i) surface modification, (ii) mechanical stirring and ultrasonication, and (iii) ball milling.

3.2.1 Surface Modification of Nanofillers

The development of polymer nanocomposites requires a compatible interface between the polymer and the nanofillers. If the interface between the nanofiller and polymer matrix is significantly different, it will lead to breakage in the final product. This generally happens when the two components (polymer and nanofiller) have a completely different chemical structure. For example, the nanofiller is hydrophilic in nature while the polymer is hydrophobic, or vice versa. In such cases, different compatibilization strategies need to be followed. One of the approaches is the chemical modification of the nanofiller surface while another approach is the use of compatibilizer additives or coupling agents. For example, montmorillonites are clay minerals having 2:1 sheet-structure: an octahedral sheet of alumina is sandwiched between two silica tetrahedral sheets, sharing an oxygen atom. These three sheets form a single clay layer called tactoid, with a thickness of approximately one nanometre. Each tactoid of the clay is separated from its neighbours by a van der Waals gap, called a gallery or interlayer, which is usually occupied by cations such as sodium

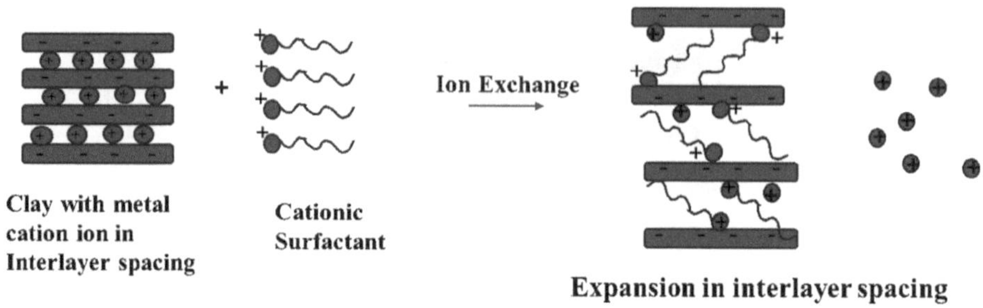

FIGURE 3.2 Schematic diagram for ion exchange reaction and enhancement of interlayer spacing of clay.

or potassium ions. Pristine-layered silicates are hydrophilic in nature and incompatible with hydrophobic polymers such as polypropylene. Therefore, pristine clay is generally modified with organic cations or surfactants to make it compatible with hydrophobic polymers. These clays are known as organic-modified clays. Moreover, these organic cations enhance wettability by lowering the surface energy. Since these cations arranged in the interlayers are generally longer and/or bigger than the original cations, they increase the gap between the layers and facilitate the penetration of polymer molecules (Figure 3.2). Furthermore, alkylammonium or alkylphosphonium groups can be linked to functional groups that interact with the polymer or promote the polymerization of the monomer. This results in an increase in interfacial resistance between the tactoids and the polymer, leading to an improvement in the final characteristics of the nanocomposite (Zhua et al. 2019).

3.2.2 Ultrasonication

The nanofiller can be exfoliated in liquid media using mechanical stirring and ultrasonication. Ultrasonication is a more effective method for exfoliating the nanofiller before solution mixing, which is also reported before melt mixing. Sound waves with frequencies more than 20kHz known as ultrasonic waves, are used to agitate the nanoparticles in solutions. Ultrasonification generates high-pressure and low-pressure regions in the liquid. In the low-pressure regions, high-intensity ultrasonic waves create small vacuum bubbles that grow and collapse, producing high-energy jets. The high-energy jets break the agglomeration or staking of particles and produce exfoliated nanofillers dispersed in the liquid. Two different types of machines are used: probe sonicators (direct ultrasonication), which are directly immersed into the polymer and nanoparticle solution and bath sonicators (indirect ultrasonication). Probe sonicators are particularly effective in utilizing the cavitation effect of ultrasonics. The major limitation of probe sonicators is the contamination of mixing solution as metal erodes from the tip of the probe. In bath sonicators, the sample container is placed in a liquid-filled bath. The ultrasonic energy is applied to this liquid. The intensity of the ultrasound wave is lower compared to that of a probe sonicator (Soman et al. 2023). Figure 3.3 shows the various steps involved in mixing of polymer and filler using ultrasonicators.

STEP 1-: Preparation of the Sample

- Weigh the NPs (e.g., silica, titanium oxide, etc.) & add them to the solvent in the beaker or the sample container

STEP 2-: Set up the Ultrasonicator

- For **probe sonicator:** Place the probe sonicator directly into the sample
- For **bath sonicator**: Place sample container into the bath filled with water.

STEP 3-: Temperature Control (optional):

- Ultrasonication generate heat, especially, during long duration. If heat could affect your sample, place the container in an ice bath.

STEP 4-: Sonication:

For **probe sonicator:**

- Start with low power settings (20-40%) and sonicate for short intervals (e.g., 1-5 minutes) to avoid overheating.

For **bath sonicator**:

- Set the sonication time (typically larger then probe sonicators, e.g., 10-30 minutes)

STEP 5-: Post -Sonication:

- After sonication, allow the sample to cool if it has been heated.
- Inspect the sample to ensure uniform dispersion. Use optical microscopy or particle size analysis to confirm dispersion quality.

FIGURE 3.3 Steps for ultrasonication process.

3.2.3 Ball Milling

Ball milling is a mechanical process that is used to grind the powder to form fine particles. A ball milling machine consists of a hollow cylinder with corrosion-resistant coating. The cylinder can be placed in a vertical or horizontal direction and rotates on its own axis. The cylinder is partially filled with grinding balls, which can be made of metal or ceramic. During the grinding process, the cylinder is filled with the nanofiller and grinding balls. As the cylinder rotates, the ball starts to grind the nanofiller

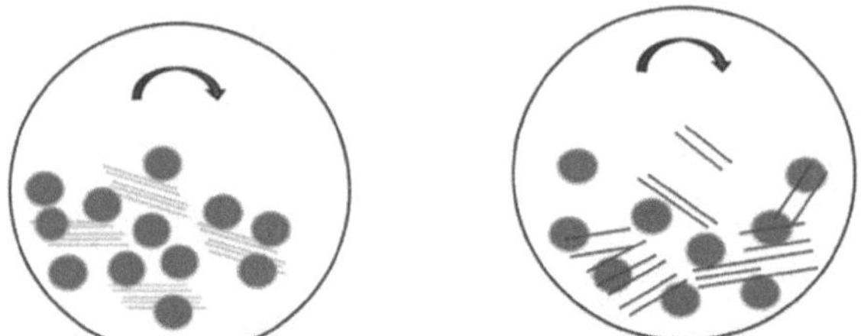

FIGURE 3.4 Schematic diagram for ball milling of nanofillers.

through attrition. When the rotational speed increases, the balls are lifted along the cylinder wall. Due to its weight, they fall back onto the particles, breaking them apart through impact force. It has been observed that two-dimensional nanofillers, such as nanoclay and graphene, are effectively exfoliated during high-energy ball milling process by both shear and impact forces (Ramadan et al. 2010; Delogu et al. 2017). The layers of the two-dimensional nanofillers are peeled off during high-energy milling process (Figure 3.4). The speed of the cylinder, ball size, the number of balls, the ball-to-nanofiller ratio, and the milling duration are some of the important process parameters for the exfoliation of nanofillers. The processing of nanoclay during high-energy ball milling changes its structure. When planetary ball milling was used to grind the kaolinite clay for 10 hours, complete delamination of the clay platelets was observed due to the breaking of hydrogen bonds between the layers. Chatterjee et al (2017) investigated the exfoliation of organo-modified clay under various ball milling conditions. The clay layers are peeled off in 2 hours at 400 rpm. When ball milling of kaolinite was carried out with formamide, it was observed that formamide intercalated between the clay layers and expended the interlayer spacing. Various alkylammonium ions have been intercalated in the presence of water and kerosene in clay, further enhancing the interlayer spacing. The graphene sheets have been exfoliated from bulk graphite, with high-energy wet milling using 2-ethylhexanol and kerosene solvents.

3.3 COMPOUNDING OF NANOFILLERS WITH POLYMER MATRIX

The nanofillers are combined with the polymer matrix in four major ways: (i) in-situ polymerization, (ii) solution mixing, (iii) melting mixing, and (iv) in-situ template synthesis of nanofillers.

3.3.1 In-Situ Polymerization

In-situ is a Latin phrase that literally translates to 'in position'. In-situ polymerization technique is a method for preparing polymer nanocomposites by incorporating and dispersing fillers in a monomer/monomer mixture, followed by chemical reactions

FIGURE 3.5 Water-based polymer nanocomposite coating using in-situ solution mixing process.

involved in polymerization (Mrah and Meghabar 2021). The dispersion of nanoparticles is the most critical and difficult parameter to achieve. Typically, mechanical dispersion or ultrasonication is used for this purpose. Polymerization is initiated by the addition of an initiator to the monomer and filler dispersion under suitable process conditions. Once polymerization is complete, it results in a nanocomposite. As an example, a water-based polymer/clay nanocomposite coating was prepared by the in-situ polymerization method, as shown in Figure 3.5. First, bentonite clay was soaked in water overnight and then ultrasonicated for approximate 1 hour. This sonicated clay dispersion was added to an aqueous solution of a monomer mixture consisting of methyl methacrylate, butyl acrylate, acrylic acid, and methacrylic acid. The clay and monomer mixture was stirred using a high-speed mixer. The initiator, ammonium persulphate, was then added to start free radical polymerization at 30°C (Sachan and Purwar 2022). The resulting nanocomposite solution was used for coating applications.

Numerous nanoparticles and monomers are used in in-situ polymerization to produce nanocomposites. Common polymers used for this technique are polystyrene, polymethyl methacrylate, polyolefins, polyurethane, polyaniline, epoxy, polylactic acid, polycaprolactone, and their copolymers. The nanoparticles that are incorporated for in-situ polymerization include clays, graphene and its derivatives, carbon nanotubes, metal oxides, and metal nanoparticles. The major advantage of in-situ polymerization is that nanofillers are uniformly distributed in the polymer matrix and that there is a strong bonding at the filler polymer surface. In addition, high concentrations of filler can be added to the polymer matrix, which can later be used as a masterbatch for the fabrication of various articles.

3.3.2 Solution Mixing

In this method, filler nanoparticles are swollen and dispersed in the polymer solution. Mechanical stirring or ultrasonication helps in the mixing and dispersion of nanoparticles in the polymer solution (Figure 3.6). The nanocomposite is then formed by either evaporation or coagulation of the solvent.

The major solvents used for the dissolution of polymers are chloroform, water, formic acid, dichloromethane, ethyl alcohol, and acetone. The main challenges of this method are cost, toxic impacts, and solvent removal. For example, silk fibroin/metal oxide nanocomposite films are produced using this method. Initially, silk fibroin was dissolved in a formic acid/$Cacl_2$ solution. Zinc oxide nanoparticles are dispersed in the solution with the help of a mechanical stirrer, followed by ultrasonication for 30 minutes (Yadav and Purwar 2021).

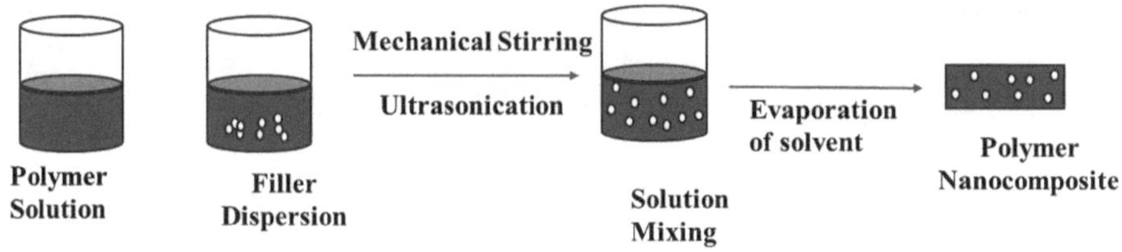

FIGURE 3.6 Schematic diagram for the preparation of nanocomposites using solution mixing process.

3.3.3 Melt Mixing

Melt mixing is mainly used for thermoplastic polymers that form stable melts, such as polyethylene, polypropylene, styrene, polycarbonate, polyamide, and so on. In this mixing process, polymer pellets and nanofillers, both in dry form, are fed into a mixing unit to produce a masterbatch. The machines typically used for melt mixing are twin-screw extruders, internal batch mixers, two roll mills, and three roll mills (Figure 3.7). The master batch is then used to fabricate articles through extrusion or moulding. In these machines, the molten polymer and nanofiller are mixed under high shear conditions (Lertwimolnun and Vergnes 2006; Pötschke et al. 2013). The incorporation of nanoparticles into the molten polymer affects the viscosity of the melt and can sometimes leads to degradation of the polymer/filler, especially when fillers are organically modified.

When clay-based nanocomposites are prepared using a twin-screw extruder, the polymer chains can enter the interlayer spaces between clay platelets due to the shearing action of screws. This results in the formation of intercalated or exfoliated structures. The extent of clay dispersion in the polymer matrix depends on the properties of the components, composition, chemical affinity between the polymer and clay, and processing conditions. Process parameters such as extruder size, extrusion rate, screw speed, screw configuration, profile temperature of barrel zones, feeding function, effective residence time, and residence time distribution affect the dispersion of nanofillers in the polymer matrix in a twin-screw extruder (Yadav and Purwar 2021; Albdiry 2024; Treece et al. 2007). Therefore, optimization of process parameters is important for uniform dispersion of nanoparticles in the polymer matrix. The major advantage of melt mixing is its economic and environmentally safe process as compared to solution mixing. Melt mixing is a simple process that enables high-speed production of nanocomposites. Therefore, this method has been widely accepted for large-scale industrial production. However, most research studies have shown that this method produces nanocomposites with partially exfoliated or intercalated structures.

3.3.4 In-Situ Template Synthesis of Nanofiller

This process is appropriate for preparing metal or metal oxide polymer nanocomposites with a narrow particle size distribution and morphology. It mainly involves two steps in the construction of nanocomposites. First, the monomer is mixed with

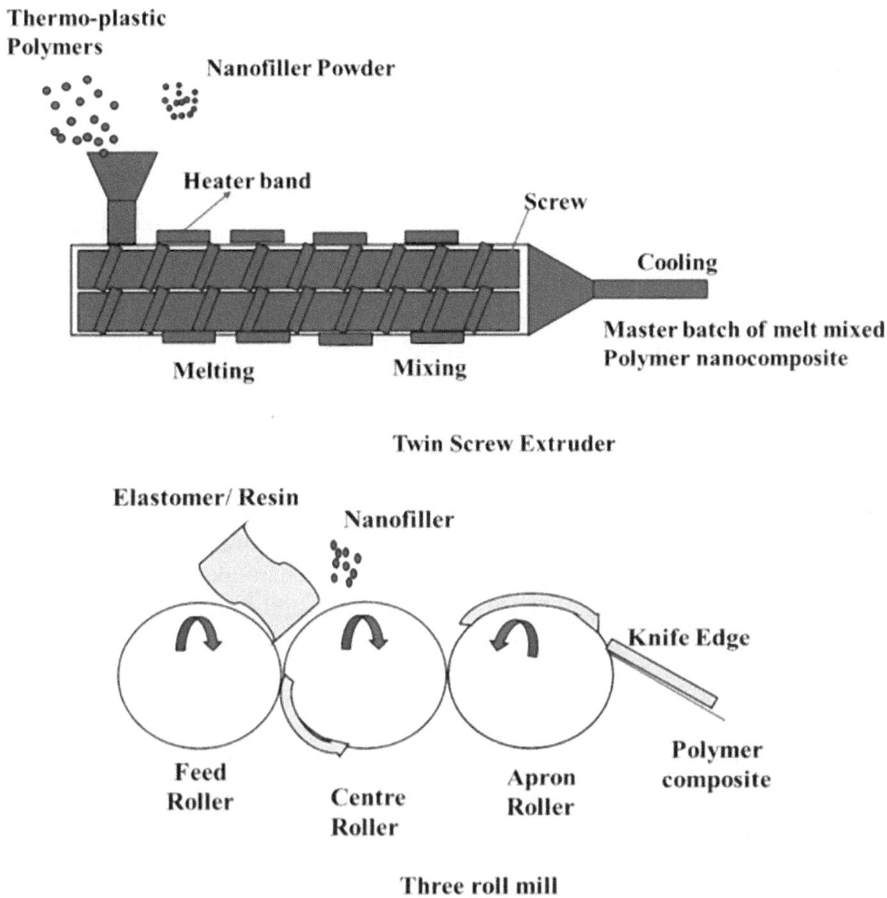

FIGURE 3.7 Melt mixing on (a) twin screw extruder and (b) three roll mill.

the precursor. Then, the monomer is converted into the polymer, and the reduction of the metal precursor into nanoparticles is carried out by a chemical reduction method. The major advantage of this method is that it produces small-sized nanoparticles that are evenly distributed in the polymer matrix. The sol-gel process is the most widely used technique for the preparation of in-situ polymer/metal or metal oxide nanocomposites (Zahran and Marei 2019).

3.3.5 Sol-Gel Process

The sol-gel process is one of the most common techniques for nanocomposite preparation in which a solution (sol) undergoes a phase transition from a liquid to a solid (gel). The flexibility and low temperature in this technique ensure that nanocomposites are prepared with a very high level of homogeneity and a controlled nanostructure. It is particularly suitable for synthesizing metal-oxide-based nanocomposites. The process of sol-gel consists of several steps, which starts with the preparation of a precursor

FIGURE 3.8 Flow chart of the sol-gel process.

solution that usually includes the dissolution of metal alkoxides or metal salts in a solvent, such as alcohol, until it becomes a transparent solution. The process continues with hydrolysis and condensation: water is added to hydrolyze the metal alkoxide, creating hydroxyl groups, which now condense into three-dimensional metal-oxygen-metal bonds to form a colloidal suspension known as a sol. It undergoes gelation over a period of time. At the end of this process, sol transforms into a gel-like network in which the solid and liquid phases coexist. Following gelation, the gel is aged and dried to enhance its structural strength, followed by the stepwise removal of the solvent through shrinkage and solidification. In specific applications, the dried gel may undergo calcination, a thermal treatment that converts the gel into a crystalline nanocomposite material while removing organic compounds completely. Incorporation of doping with other materials may be present during sol preparation for introducing secondary phases or significantly enhance some properties of the nanocomposite in most applications. This synthesis method offers high control over elemental composition and material structure, making it a widely used technique for synthesizing advanced nanocomposites. A block diagram outlines the sol-gel process, showing the stages from the precursor solution to the final nanocomposite (Figure 3.8).

For example, a ZnO/ chitosan polymer nanocomposite is prepared using the sol-gel method. The synthesis begins with zinc acetate as the precursor for ZnO. The zinc acetate is dissolved in water, and hydrolysis is initiated by adding a base, such as NaOH, to form ZnO nanoparticles. In parallel, chitosan is dissolved in an acidic solution. Once both solutions are prepared, the ZnO sol is gradually added to the chitosan solution under continuous stirring, allowing for the integration of ZnO into the chitosan matrix. This mixture is stirred until gelation occurs, forming a solid network. After gelation, the gel is aged to strengthen its structure, dried to remove excess water, and finally calcined to improve crystallinity, resulting in a ZnO/chitosan nanocomposite with strong antibacterial properties (Wu and Li 2024).

3.4 POLYMER NANOCOMPOSITE BY ELECTROSPINNING TECHNIQUE

Electrospinning is a simple electrohydrodynamic process in which a polymer jet is generated by applying high voltage to a polymer solution droplet. The droplet is stretched and elongated under high electrical field. The diameter of the fibres obtained through this process ranges from 500 nm to 10 nm. Figure 3.9 illustrates the basic setup for an electro spinning unit. It consists of a high voltage power supply, syringe pump, syringe which acts as the spinneret and polymer solution dope supplier, and collector plate or cylinder. A polymer solution is prepared in a suitable solvent for electrospinning and fed into the syringe. The syringe is then connected

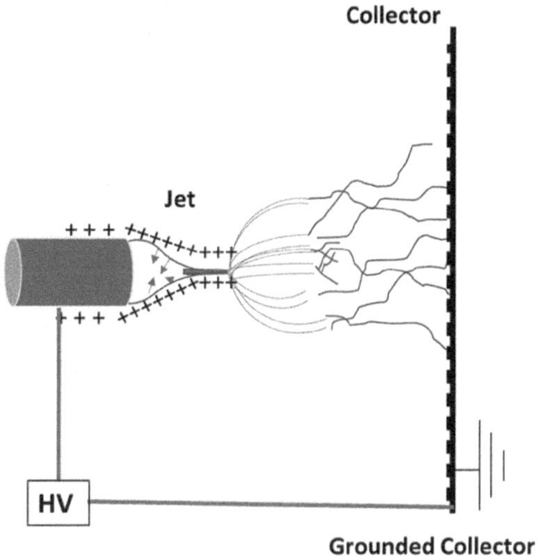

FIGURE 3.9 Schematic diagram for the electrospinning process.

to a high voltage source, which applies the electric field and generates an electric charge over it. A charged jet is produced from the syringe when the surface tension is overcome by the repulsive electric forces, leading to solvent evaporation and the fibres spun are collected on the rotating collector plate. During electrospinning, the polymer solution is prepared in a suitable solvent.

This solution is called as the dope solution. This dope solution is fed into a syringe and the syringe tip is connected to a high-voltage power supply. Initially, the pump supplies the dope solution to the syringe tip, forming a droplet due to surface tension. When voltage is applied, the tip and the droplet acquires the same sign (positive or negative) and further increases in voltage deform into a cone (Taylor cone). The polymer dope solution is ejected in the form of a jet, which initially extends as a straight line before undergoing dynamic whipping motions caused by bending instabilities. The liquid jet stretches into a fine line. As the solvent evaporates, it is converted into a solid and deposited on the collector in the form of fine fibres. These fine fibres form a nonwoven structure. The electrospinning process is influenced by various machine parameters, including voltage, flow rate, tip to collector distance, and the type of collector. Material parameters that affect the properties of fibres are concentration, viscosity, conductivity, surface tension, and molecular weight of the polymer. Ambient experimental conditions such as temperature and humidity also affect the properties. Electrospun nanofibrous nonwovens have smaller diameters, higher porosity, large surface area to volume ratio, interconnected porous structures, tuneable network geometry, and good internal connectivity (Purwar 2013, 2020; Bansal and Purwar 2023).

Sericin/PVA/Cloisite 30B nanoclay nanofibrous mats are prepared using an electrospinning machine. The dope solution is prepared in distilled water by dissolving equal amounts (10% wt/wt) of Sericin and PVA at 80°C. The dope solution for Sericin/PVA/Clay is prepared by adding different clay concentrations ranging from

0.1% to 0.75% (% of polymer) and subjected to ultrasonication. The process parameters for the electrospinning machine are varied as follows: acceleration voltage (18–28 kV), spinning distance (3–12 cm), and flow rate (0.8–1.2 mL/hour). The collector is covered with a nylon net fabric to collect the Sericin/PVA/clay nanofibrous mat. It was observed that the nanoclay inside the nanofibrous nanocomposite became fully exfoliated. Cloisite 30B contains $NH4^{+}$ ions in its interspace gallery. During electrospinning, a high positive charge is applied at the nozzle, which gives positive charge to the outer layer of the dope solution. This high positive charge repels the cations present in the clay which may lead to breakdown of the stacked structure of clay. The high electric field will also align the clay platelets. The collector plate having a negative charge will attract these clay platelets, resulting in an exfoliated Sericin/PVA/Clay nanocomposite structure (Purwar et al. 2013).

3.5 POLYMER NANOCOMPOSITE BY 3D PRINTING

3D printing, or additive manufacturing, is a process that initiates and builds objects layer by layer from digital models. It can manufacture intricate structures with high accuracy. In the context of nanocomposites, it allows for the preparation of polymer matrices doped with nanoparticles, thus offering improved mechanical, thermal, electrical, or biological features. The process begins with the selection of suitable precursors and the combination of a matrix material with nanoparticles, such as polymers, ceramics, metals, carbon nanotubes, graphene, or metal oxides. Then, nanofillers (nanoparticles, nanotubes, or fibres) are dispersed in the selected matrix material. The dispersion methods vary with the type of materials used: the sol-gel process, polymerization, and melt blending. It is essential to have an evenly dispersed nanofiller for the final product to ensure consistent similar properties. The synthesized nanocomposite material is then used in additive manufacturing techniques such as fused deposition modelling (FDM). Generally, the post-processing treatments that the 3D structure undergoes are diversified and include annealing, curing, or surface modifications. Such treatments ensure that the bond between the matrix and nanoparticles becomes more robust so that the mechanical properties of the final product are improved (Liu and Zhu 2023; Rashid et al. 2021, and Mandala et al. 2022). Figure 3.10 shows the block diagram for preparation of nanocomposites using 3D printing process.

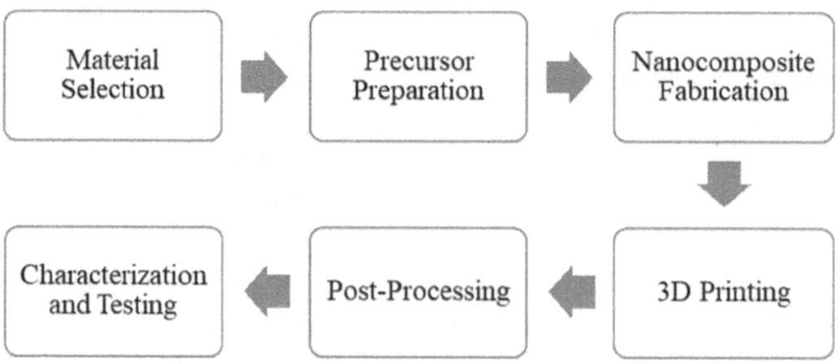

FIGURE 3.10 Block diagram for the preparation of nanocomposites using 3D printing.

3.6 FILAMENT-BASED 3D PRINTING (FUSED DEPOSITION MODELLING)

In this process, an object is developed by extruding melted plastic through a heated nozzle and depositing it layer after layer. Nanocomposites may be added by mixing nanoparticles—for example, carbon nanotubes, graphene, or metal-oxide nanoparticles—into thermoplastic filaments such as PLA (polylactic acid), ABS (acrylonitrile butadiene styrene) or PETG (polyethylene terephthalate glycol). The process begins by loading a spool of thermoplastic filament (PLA or ABS). Next, the filament is moved through a heated extruder, where it melts to prepare for the extrusion process. The printer nozzle moves along the X, Y, and Z directions, depositing the molten filament onto the build platform according to the design of every layer. Once each layer is laid down, the platform lowers in preparation for adding the next layer. As the molten filament cools and solidifies, it bonds to the previous layer to build the object's structure. This process of layer deposition, cooling, and repetition continues until the 3D object is printed. The precision of how each nozzle moves, the flow of filaments, and cooling contribute toward having an accurate and robust printed object. The major process parameters are nozzle temperature, speed of printing, thickness of the layer, infill density, and bed temperature. An increased nozzle temperature increases the filament flow. It makes the flow more quickly, but it begins to degrade the polymer material and introduces defects, such as warping. Faster printing speeds increase productivity but reduce the print quality, especially in terms of layer bonding. A balance must be achieved to ensure proper layer bonding and nanoparticle alignment in nanocomposites to ensure that the integrity of the material is not compromised. This would affect the quality of the surface and mechanical properties of the final product. Higher density would render the parts more robust but waste more material and time. A heated bed prevents warping of the base by ensuring good adhesion to the build surface, especially with ABS-type materials. Cooling fans ensure that the material solidifies quickly, improving print quality. Also, especially with PLA, excessive cooling sometimes causes poor layer bonding. Graphene-PLA nanocomposites represent significant progress in filament-based 3D printing. Graphene contributes superior mechanical, thermal, and electrical properties, while PLA provides biocompatibility and biodegradability. The process starts by preparing the nanocomposites by mixing graphene nanoplatelets or graphene oxide with PLA pellets to ensure uniform dispersion, which is essential for optimizing the intended properties. Once extruded, parameters such as temperature and cooling rate are controlled to prevent graphene agglomeration. Once the graphene-PLA is loaded into a standard FDM 3D printer, the nozzle extrudes the material layer by layer, according to the CAD model, at temperatures typically ranging from 190°C to 220°C. The improved flow characteristics of nanocomposites result in easier extrusion and better layer adhesion, which improves the mechanical strength, electrical conductivity, and thermal stability of the printed parts. These properties make graphene-PLA nanocomposites ideal for applications in aerospace, automotive lightweight structures, flexible electronics, and biomedical devices such as drug delivery systems and tissue scaffolds. By embedding graphene into PLA using filament-based 3D printing, advanced materials aim to solve complex challenges and create innovative solutions across many different sectors.

3.7 SOLUTION-BASED 3D PRINTING (DIRECT INK WRITING – DIW)

In this process, a viscous solution or gel containing the material (polymer + nanoparticles) is extruded through a nozzle and immediately solidifies at that location to create the desired shape. This technique enables complex architectures with controlled material composition and spatial distribution, making it highly suitable for hydrogels, ceramics, and bio-based materials. The process starts by loading the material into a syringe or extrusion system in the form of a solution or slurry. This material usually consists of polymers, ceramics, biomaterials, and nanocomposites dissolved or dispersed in a solvent or matrix. A pressure-driven extrusion mechanism is used to push the material through the nozzle, which moves in the X, Y, and Z directions and accurately deposits the material based on the object's CAD model. After each layer is deposited, the extruded material undergoes solidification due to solvent evaporation or chemical cross-linking, such as UV or thermal curing, or drying over time, ensuring it retains its shape. The gelling agent hardens the printed structure by acting as a hydrogel solidifier. Layer-by-layer deposition, along with the consolidation of the printed structure, is repeated until the complete object is established, ensuring the printed structure's precision and stability. The major process parameters are ink viscosity, extrusion pressure, and drying rate. The ink must be viscous enough to retain its shape after being deposited but not so thick that it causes clogging in the nozzle. Higher extrusion pressures would be required for depositing more viscous inks; otherwise, high extrusion pressures can lead to inaccuracies in the printed structure. Nanocomposite inks should have a shear-thinning behaviour to facilitate the flow effortlessly during extrusion but should be solid once deposited. Water-based systems must control the drying rate to prevent shrinkage-related defects that occur during drying.

3.8 CHALLENGES AND FUTURE PERSPECTIVE

A significant challenge in the processing of nanocomposites is related to the scale-up of production. While laboratory-scale synthesis methods such as in-situ polymerization, melt mixing, solution mixing, and sol-gel processes can produce high-quality nanocomposites, replicating these results at an industrial scale can be problematic. Issues such as ensuring consistent quality, controlling the processing environment, and managing the cost of production are all factors that need to be addressed during scaling up. Moreover, health and safety concerns associated with the handling of nanomaterials must be considered, as their small size allows them to penetrate biological barriers, posing potential risks to workers and the environment.

Despite these challenges, the potential applications of polymer nanocomposites are vast and varied, ranging from packaging and automotive components to biomedical devices and electronic systems. The enhanced properties that nanocomposites offer, such as improved strength, flame retardance, and thermal stability, make them highly desirable for these applications. As progresses, new methods and technologies are being developed to overcome the existing challenges. This advancement promises to expand the use of polymer nanocomposites across various industries and open up new possibilities for innovative materials and products. The ongoing exploration of polymer nanocomposites is a testament to the dynamic nature of materials science and its ability to push the boundaries of what is possible with modern engineering and technology.

REFERENCES

Albdiry, M. (2024) Effect of melt blending processing on mechanical properties of polymer nanocomposites: A review. *Polym. Bull.* 81:5793–5821

Bansal, P., Purwar, R. (2023) "Nanofibers' utility for rejuvenation of heavily contaminated environments," edited by Charles, O.A., Singh, R.P., Singh, J., Kshitij, R.B., *Nanobiotechnology for Bioremediation* (pp. 23–50). Elsevier.

Chatterjee, U, Butola, B.S., Joshi, M. (2017) High energy ball milling for the processing of organo-montmorillonite in bulk. *Appl. Clay Sci.* 140:10–13.

Delogu, F., Gorrasi, G., Sorrentino, A. (2017) Fabrication of polymer nanocomposites via ball milling: Present status and future perspectives. *Prog. Mater. Sci.* 83:75–123.

Joshi, M., Adak, B., Butola, B.S. (2018) Polyurethane nanocomposite-based gas barrier films, membranes and coatings: A review on synthesis, characterization and potential applications. *Prog. Mater. Sci.* 97:230–282.

Lertwimolnun, W., Vergnes, B.(2006) Effect of processing conditions on the formation of polypropylene/organoclay nanocomposites in a twin screw extruder, *Polym. Eng. Sci.* 43:314–323

Liu, C., Zhu, W. (2023) 3D printing of graphene-reinforced PLA nanocomposites: Mechanical properties and thermal behavior. *J. Appl. Polym. Sci.* 140(12):345–357.

Mandala, R., Bannoth, A.P., Akella, S., Rangari, V.K., Kodali, D., (2022) A short review on fused deposition modeling 3D printing of bio-based polymer nanocomposites. *J. Appl. Polym. Sci.* 139:51904.

Mrah, L., Meghabar, R. (2021) *In situ* polymerization of styrene–clay nanocomposites and their properties. *Polym. Bull.* 78:3509–3523.

Peponi, L., Puglia, D., Torre, L., Valentini, L., Kenny, J.M. (2014) Processing of nanostructured polymers and advanced polymeric based nanocomposites. *Mater. Sci. Eng. R.* 85:1–43.

Potts, J.R., Dreyer, D.R., Bielawski, C.W., Ruoff, R.S. (2011) Graphene-based polymer nanocomposites. *Polymer* 52:5–25.

Pötschke, P., Krause, B., Buschhorn, S.T., Köpke, U., Müller, M.T., Villmow, T., Schulte, K. (2013) Improvement of carbon nanotube dispersion in thermoplastic composites using a three roll mill at elevated temperatures. *Compos. Sci. Technol.* 74:78–84.

Purwar, R. (2020) "Bionanocomposites Based on Silk Proteins and Nanoclay," edited by Mangala J., *Nanotechnology in Textiles: Advances and Developments in Polymer Nanocomposites*. Pan Stanford Publishing (pp.179–204).

Yadav, R., Purwar, R. (2021) Influence of metal oxide nanoparticles on morphological, structural, rheological and conductive properties of mulberry silk fibroin nanocomposite solutions. *Polym. Test.* 93:103913.

Purwar, R., Goutham, K.S., Srivastava, C.M. (2013) Electrospun sericin/PVA/clay nanofibrous mats for antimicrobial air filtration mask. *Fibers Polym.* 17:1203–1213.

Ramadan, A.R., Esawi, A.M.K., Gawad, A.A. (2010) Effect of ball milling on the structure of Na+-montmorillonite and organo-montmorillonite (Cloisite 30B). *Appl. Clay Sci.* 47:193–202.

Rashid, A.A., Khan, S.A. Al-Ghamdi, S.M., Koç, M. (2021) Additive manufacturing of polymer nanocomposites: Needs and challenges in materials, processes, and applications. *J. Mater. Res. Technol.* 14:910–941

Sachan, R., Purwar, R. (2022) Water based quad acrylic copolymer/cloisite 30B nanocomposite heat resistant adhesive: thermal and rheological properties. *J. Adhes. Sci. Technol.* 33:1541–1553.

Soman, V., Vishwakarma, K., Poddar, M.K. (2023) Ultrasound assisted synthesis of polymer nanocomposites: A review. *J. Polym. Res.* 30: 403.

Treece, M. A., Zhang, W., Moffitt, R. D., Oberhause, J. P. (2007) Twin-screw extrusion of polypropylene-clay nanocomposites: Influence of masterbatch processing, screw rotation mode, and sequence. *Polym. Eng. Sci.* 47:898–911.

Wu, Y., Li, Y. (2024) Green synthesis of ZnO/chitosan nanocomposite using the sol-gel method for biomedical applications. *J. Nanobiotechnol.* 22:123–135.

Zahran, M., Marei, A. H. (2019) Innovative natural polymer metal nanocomposites and their antimicrobial activity. *Int. J. Bio. Macromol.* 133:583–593.

Zhanga, W., Camino, G., Yanga, R. (2017) Polymer/polyhedral oligomeric silsesquioxane (POSS), nanocomposites: An overview of fire retardance. *Prog. Polym. Sci.* 37:77–125.

Zhua, T.T., Zhoua, C.H., Kabwe, F.B., Wua, Q.Q., Lid, C.S., Zhanga, J.R. (2019) Exfoliation of montmorillonite and related properties of clay/polymer nanocomposites. *Appl. Clay Sci.* 139:48–33.

4 Enhancement of Mechanical Properties in Polymer Nanocomposites

Deepa Sharma and Purnima Jain

4.1 INTRODUCTION

Nanomaterials are an attractive class of advanced materials that are governed by their structural and dimensional properties for various applications. They are known for their unique mechanical, optical, magnetic, chemical, electrical, and other properties. These properties have a great impact and extend their applications in sensing, energy devices, medicine, catalysis, cosmetics, electronics, and other fields (Darwish et al., 2022; Wing et al., 2003). Nanomaterials are added to polymers to make them stronger and lighter to develop smart fabrics, lightweight high-performance military aircraft, automobiles, high-performance portable batteries, solar cells, targeted drug deliveries, nanotransistors, and plasma displays (Tyagi & Tyagi, 2014; Shahbaz et al., 2023). Nanomaterials possess a large surface area and high surface activity, which make them an attractive choice for catalysis (Chadha et al., 2022), cosmetic applications (Lu et al., 2015; Vaudagna et al., 2023), manufacturing of nanofertilizers, and nanopesticides (Tang et al., 2023). Thus, incorporating nanofillers into the polymeric matrix enhances the overall characteristics of a polymer nanocomposite (PNC) than ordinary composites (Figure 4.1).

The nanoscale dimension (1–100 nm) of reinforcing nanofillers is responsible for enhancing the physicochemical properties of a nanocomposite material, such as a high aspect ratio, large surface activity, and great mechanical properties, barrier properties, scratch resistance, and fire resistance at low loadings (Shrivastava, 2018). Besides, it also aids in the enhancement of characteristic properties of nanocomposites, such as electrical, thermal, optical, and photonic properties. Recent research reported the enhancement of PVA's structural, optical, thermal, and electrical properties on doping bimetallic nanoparticles of TiO2/Cu for energy storage capacitors applications and in the alternative-rechargeable lithium-ion battery industry (Al-Hakimi et al., 2023). Another research highlights that the tensile strength and stiffness of the composite can be significantly increased using carbon nanotubes (CNTs) or graphene while also increasing the toughness of the material (Bal & Samal, 2007).

Mechanical properties are considered the most important properties of a plastic material among all the properties. This is because they help in determining the fundamental insights about the physical characteristics of a material. They are typically

DOI: 10.1201/9781003527602-4

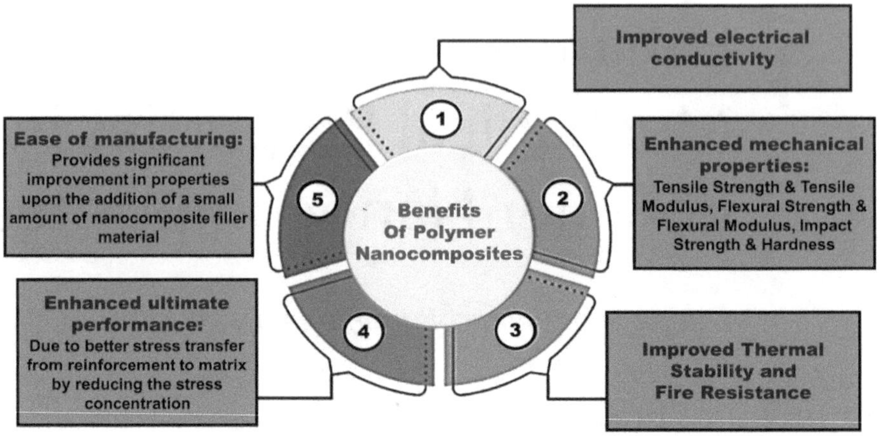

FIGURE 4.1 Benefits of polymer nanocomposites over conventional composites (Rashid et al., 2024).

the structural and dimensional characteristics of a PNC that define its practical application. Mechanical properties are crucial aspects of nanocomposites, particularly in structural applications where load-bearing capacity and durability are essential. Enhancing these properties of PNC expands their utility across a wide range of industries, including automotive, aerospace, electronics, and biomedical engineering (Ramakoti et al., 2023). PNC with modified mechanical properties is stronger, more flexible, and resilient for designing lightweight materials that could replace metals or ceramics in various applications as cost-effective and energy-efficient alternatives (Müller et al., 2017; Nan et al., 2023; Singh et al., 2023). Therefore, the material selection for any application is often based on substantial mechanical properties such as tensile strength, impact strength, toughness, hardness, and elongation.

Meesorn with his co-workers (2017) developed a strategy to improve the mechanical properties of PNCs with cellulose nanocrystals (CNCs). They found that the mechanical properties of polymer/CNC nanocomposites do not match the theoretical predictions due to the aggregation of CNCs. To avoid this aggregation, polyvinyl alcohol (PVA) was used as a binder, and with the addition of 1–5% of PVA to the poly(ethylene oxide-*co*-epichlorohydrin)/CNC nanocomposites, a remarkable change in mechanical properties was observed. There were remarkable enhancements in the stiffness and strength of the nanocomposite at PVA content as low as 1%. In addition to this, increasing PVA loading to 5% resulted in the enhancement of Young's modulus, storage modulus, and strength up to 5-fold in comparison to non-PVA-reinforced nanocomposite (Meesorn et al., 2017). The formation of PVA-incorporated poly(ethylene oxide-*co*-epichlorohydrin)/CNC nanocomposites is the best example of a ternary nanocomposite with improved mechanical properties. Additionally, the incorporation of metal nanoparticles as nanofillers facilitates mechanical property enhancement and modifications. For example, Somoghi et al. (2024) used ZnO and functionalized ZnO nanofillers as reinforcing agents within the epoxy resins to achieve suitable mechanical properties, high adhesion strength, and corrosion resistance. The final properties of ZnO/epoxy nanocomposite depend upon several factors such as the nature of ZnO nanofiller, its loading, epoxy resin type, curing agents, and the method adopted for nanocomposite formation. With enhanced mechanical properties, adhesion strength, and corrosion resistance, ZnO/epoxy-based nanocomposite could be used for various applications, including anti-corrosion and UV-protective coatings, for adhesives or use in building materials or electronics (Şomoghi et al., 2024). In conclusion, the surface functionalization of nanofillers and synergistic mixing of two or more polymeric components is an innovative way to improve the mechanical characteristics of nanomaterials.

This book chapter highlights all major aspects of the mechanical properties of a PNC and discusses their role in the development of advanced PNC. It also reviews the enhancement of mechanical properties and the factors influencing these properties. Some of the major factors include nanofillers and their dispersion, surface functionalization, and interfacial compatibility between nanofillers and polymers (bulk material). Furthermore, the processing techniques for the fabrication of PNC, such as solution blending, melt mixing, and in-situ polymerization, and their influence on characteristic mechanical properties are also covered (Kamal et al., 2022; Zotti et al., 2022). By addressing these objectives, this chapter aims to provide a comprehensive understanding of the current methodologies for enhancing the mechanical properties of PNCs, highlight the limiting challenges, and identify future research directions for further improvement and application.

4.2 OVERVIEW ON MECHANICAL PROPERTIES

Mechanical properties are defined as the behavior of polymers under stress. Mechanical properties are addressed as the structural and dimensional properties of the materials. They are the material-defining properties of polymeric material among all the properties to determine the end-user applications of PNC. They help in determining the material properties for use in the design and analysis of the product, production process, quality control, and application performance requirements. However, to understand the mechanical properties, it is important to develop a basic understanding of the stress–strain behavior of plastic materials.

Stress is defined as the ratio of force applied (load) to the original cross-sectional area (lb/in^2) of a test specimen to produce deformation. Strain is the ratio of change in length per unit of the original length (gauge length) of the test specimen, expressed as a dimensionless ratio ($\Delta l/l$), whereas elongation is the increase in length of the test specimen under a tensile load. Figure 4.2 depicts a typical stress–strain (load deformation) diagram. The initial curve from A to B is linear and follows Hooke's Law, stating that stress is proportional to strain for elastic materials. The deformations up to point B are associated with the stretching and bending of the interatomic bonds of the plastic materials. Such deformations are relatively small, instantaneous, and recoverable, showing the elastic behavior of the material. Furthermore, C is called the proportional limit, as up to this point only the stress is proportional to strain, after that, the actual curve deviates from the straight line. Above this limit, the deformations of the plastic materials are recoverable ultimately but not instantly. The yield point of a plastic material is depicted in Figure 4.2 D which is an increase in strain

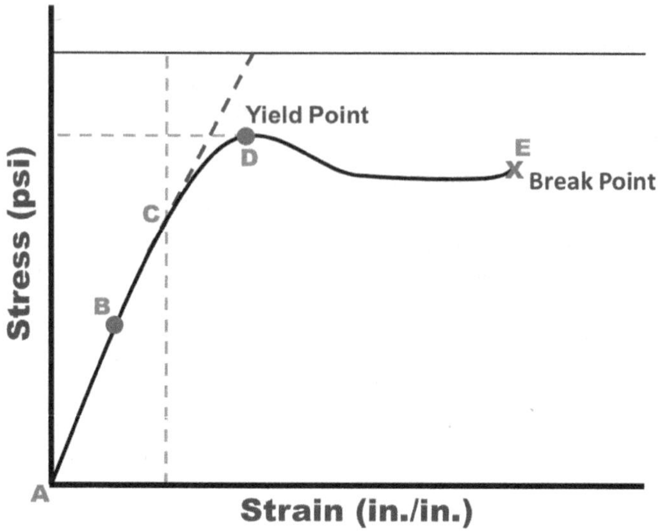

FIGURE 4.2 Illustration of a typical stress–strain curve.

without any change in stress, also termed as an elastic limit. Beyond the elastic limit, the deformations of the materials are nonrecoverable because of the actual displacement of the molecules concerning each other. Therefore, it is said to be a permanent deformation or set of plastic materials. All these deformations do not occur separately and are superimposed on each other.

These deformations can be further explained with the Maxwell model, which duplicates plastic behavior under various external conditions. Maxwell model is also known as the spring and dashpot model, as shown in Figure 4.3. The spring is assumed to exhibit normal elastic behavior, and its extension is analogous to the deformations that occur because of bending and stretching of the interatomic bonds. If a linear spring is replaced with a nonlinear one, then the deformations are similar to the uncoiling of molecular chains. The dashpot is filled with a viscous fluid that leaks through holes in the plunger disc before the disc can move. The extension in the dashpot is nonrecoverable and corresponds to permanent deformation as a result of intermolecular slippage.

Due to certain mechanical characteristics, polymeric materials can be classified in terms of their relative hardness, softness, brittleness, and toughness (Table 4.1). This could be determined by the stress–strain curves of the material, where the area under the curve provides the information regarding the toughness of the polymeric materials. This is shown in Figure 4.4.

4.2.1 Tensile Tests (ASTM D 638, ISO 527-1)

Tensile strength is the measurement of the ability of a material to withstand the force that tends to tear it apart and determines the extent to which it could stretch before breaking. Tensile modulus is the measurement of the relative stiffness of a material,

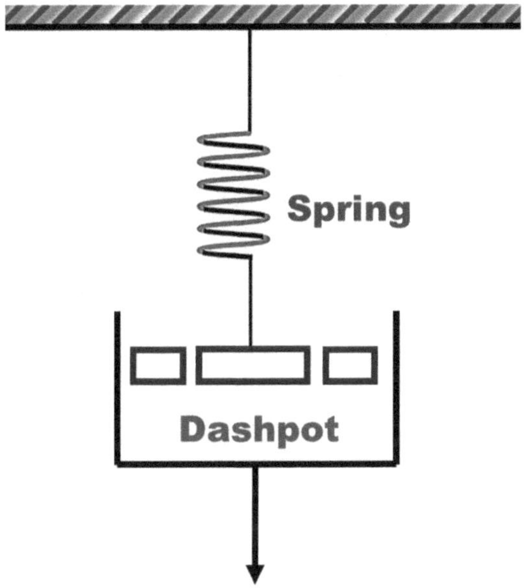

FIGURE 4.3 Maxwell model.

TABLE 4.1
Characteristic Features of Stress–Strain Curves

Nature of Polymer	Modulus	Yield Stress	Ultimate Strength	Elongation at Break
Soft, weak	Low	Low	Low	Moderate
Soft, tough	Low	Low	Yield stress	High
Hard, brittle	High	None	Moderate	Low
Hard, strong	High	High	High	Moderate
Hard, tough	High	High	High	High

which is determined from the stress–strain curve. Tensile strength and tensile modulus measurements are among the most influential factors in indicating the strength of a material and are widely specified for plastic or polymeric materials. Tensile testing simplifies the preferential selection of polymeric materials from a large group of polymers. However, the test does not take into account the time-dependent behavior of plastic materials, thus limiting their use in the actual design of the product.

The tensile strength of PNC is typically higher than that of pure polymers. This enhancement is attributed to the strong interfacial interactions between the polymer matrix and the nanoparticles, which facilitate efficient stress transfer from the polymer to the nanoparticles (Rashid et al., 2024). Nanoparticles such as CNTs and graphene possess extremely high intrinsic strength and modulus, which, when well-dispersed within the polymer matrix, contribute significantly to the overall

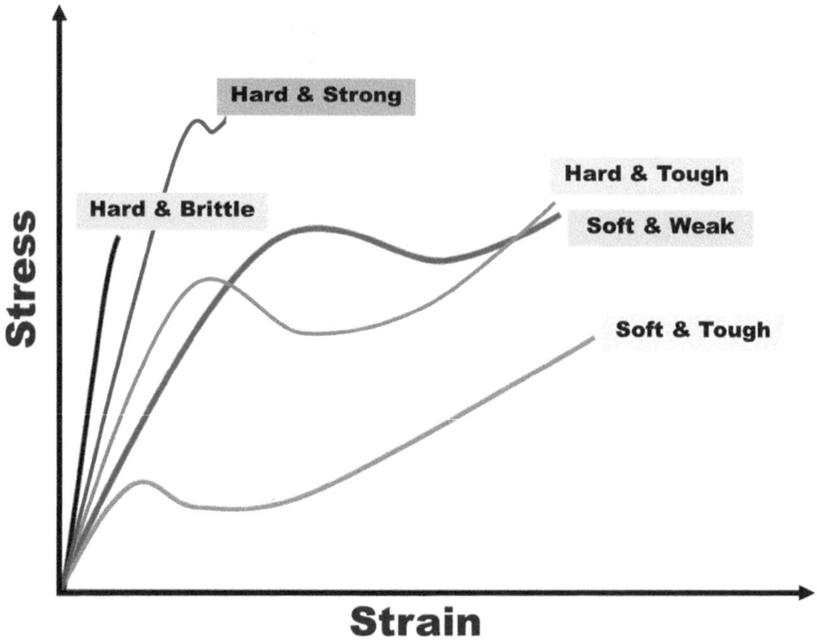

FIGURE 4.4 Different types of stress–strain curves based on variable strength of polymeric materials.

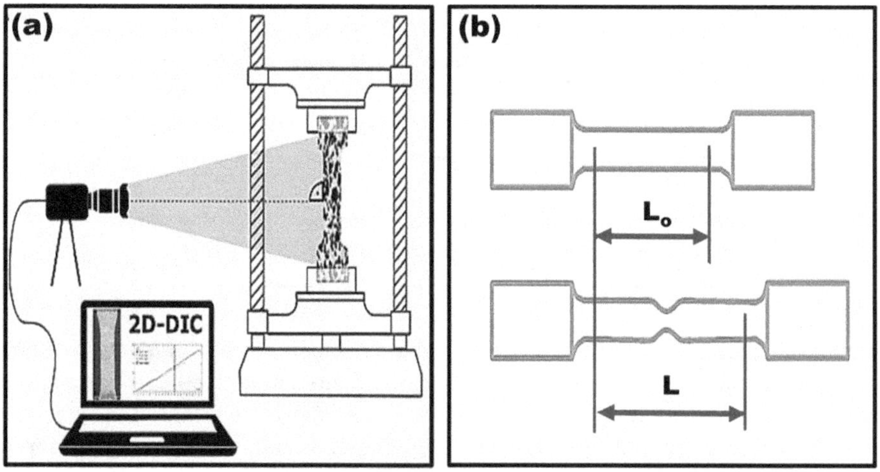

FIGURE 4.5 (a) Universal tensile testing machine and (b) standard tensile test specimen.

mechanical performance (Guan & Tang, 2021; Ma et al., 2010; Saeed et al., 2023; Song et al., 2020).

Figure 4.5 demonstrates the advanced universal testing machine (UTM) for tensile testing and standard test specimens, which are subjected to controlled tension until failure. UTM has two gripping ends, one end is fixed while another is movable,

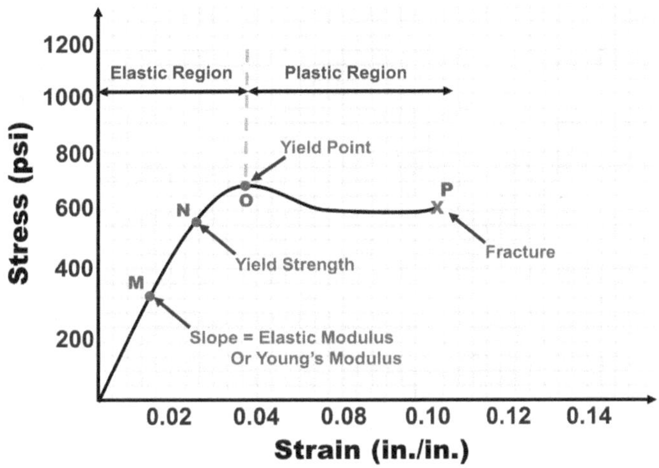

FIGURE 4.6 Illustration of a stress–strain curve from which the values of modulus and elongation are derived.

which pulls away the material when a specified load is applied. The testing machine of a constant rate-of-crosshead movement is used with a controlled-velocity drive mechanism for a high degree of speed accuracy. All the parameters, such as stress, elongation, energy, modulus, and statistical calculations, are performed automatically using system system-generated stress–strain curve as shown in Figure 4.6 and presented on a display screen, which could be printed later.

Tensile test specimens are flat, dumbbell-shaped, as shown in Figure 4.5b and are prepared either by injection molding or compression molding or machining. They may also be prepared in the form of sheets, plate, or slab from materials. The ASTM book of standards is followed for determining the dimensions of test specimens as per the requirements, such as ISO 527-4, ISO 527-5, ASTM D 638, ASTM D 3034, and ASTM C 247. ASTM D 638 Type I is commonly used for testing rigid and semirigid plastics (Saba et al., 2019). The specimens are conditioned using standard conditioning procedures, and it is recommended to conduct the test in the standard laboratory conditions of 23±2°C and 50%±5% relative humidity. Procedure A of ASTM D 618 is recommended for this test.

4.2.1.1 Factors Affecting the Tensile Test

Molecular orientation and specimen preparation: The molecules that are aligned parallel to the direction of the load applied yield higher tensile strength values than the ones oriented perpendicular to the load. However, it is the opposite in the case of tensile modulus. Figure 4.7 illustrates an example of glass fibers (nanofillers) and the effect of fiber orientation on the tensile strength. The method employed for the preparation of specimens also plays a significant role. The test specimen prepared using injection molding generally yields high strength than compression-molded specimens. Machined test specimens have generally lowered tensile and elongation values due to small irregularities introduced during machining. Also, the location and size of the gate on molded specimens similarly affect these values as the orientation of nanofillers.

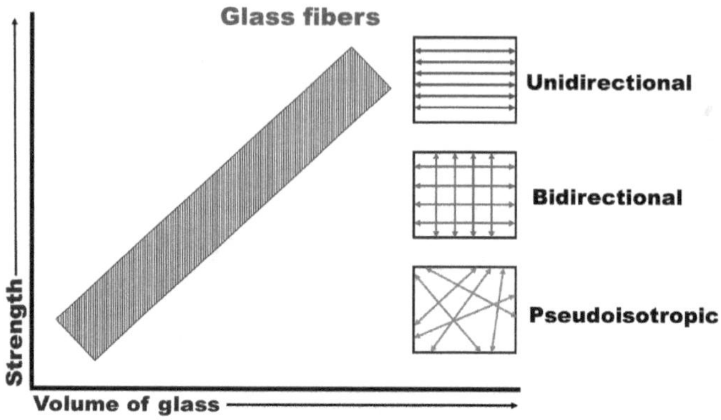

FIGURE 4.7 Effect of the orientation of nanofillers (glass fibers) on the strength.

A large gate on top of the tensile bar will orient parallel to the applied load and increase the tensile strength value, while a gate located on the side of the specimen would disperse the nanofillers in random orientation, thereby reducing the strength values.

Strain rate: It is defined as the rate at which strain (deformation per unit length) changes with respect to time. It typically measures how quickly a material deforms under stress during testing. Thus, the tensile strength and modulus of the test specimen increase with an increase in strain rate. However, the strain rate is inversely proportional to the elongation.

Temperature: It principally governs the tensile properties of a material. Some polymeric materials might become more pliable at high temperatures and more brittle at lower temperatures. Thereby, the tensile properties of some materials change rapidly with small temperature changes. The tensile strength and modulus are decreased while the elongation is increased with temperature.

4.2.2 Flexural Properties (ASTM D 740, ISO 178)

Flexural strength is the ability of a material to withstand bending forces applied perpendicular to its longitudinal axis. Flexural tests, conducted using three-point or four-point bending setups according to **ISO 14,125, ISO 178, ASTM D 740, and ASTM D 6272** testing standards, measure the ability of a material to resist deformation under load (Saba et al., 2019). The three-point loading system involves center loading on a simple supported beam as shown in Figure 4.8. This flexural testing method is suitable for materials that break at small deflections, while method 2 (four-point bending) is suitable for large deflections. In the latter, two load points are placed equally spaced from their adjacent support points as shown in Figure 4.9. The maximum axial fiber stress occurs over the area between the loading noses. This method is for materials that fail under a three-point system under maximum stress. The basic difference between the two methods is their strain rate; three-point systems have 0.01 in./in./min strain rate while four-point systems have 0.10 in./in./min.

The stress induced during flexural load consists of compressive and tensile stresses (Figure 4.8). Thus, flexural properties are reported and calculated as the maximum

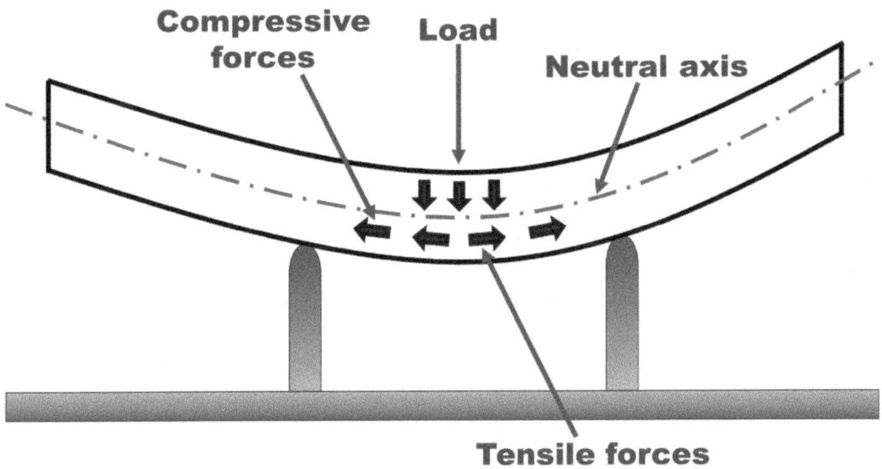

FIGURE 4.8 Diagrammatic representation of forces involved in the bending of a test specimen during three-point loading system of flexural testing.

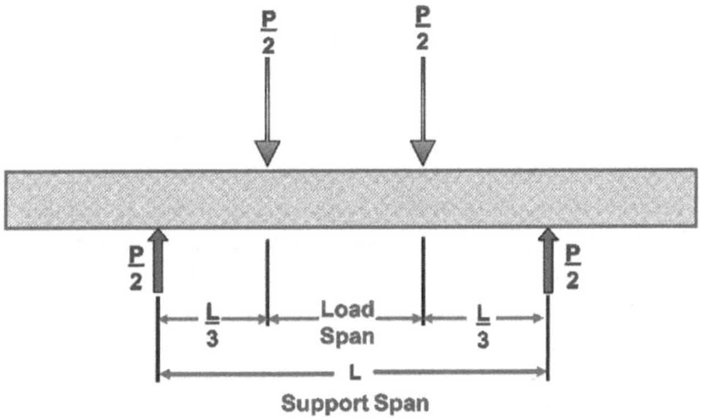

FIGURE 4.9 Schematic for four-point flexural testing system.

stress and strain experienced on the outer surface of a test material. The flexural strength is the stress at failure in bending, while the flexural modulus is the slope of the stress–strain curve in the elastic deformation region (ASTM D740-10).

4.2.2.1 Factors Affecting the Flexural Test

Molecular orientation and specimen preparation: The material with a high degree of molecular orientation perpendicular to the applied load will show higher flexural strength and modulus than those with parallel orientation of molecules against the load. Injection-molded specimens show higher flexural values than compression-molded test specimens.

Temperature: The flexural strength and modulus values are inversely proportional to the temperature.

Test conditions: The strain rate plays a key role during flexural testing as it depends upon the testing speed, specimen thickness, and the distance between the supports (span). The flexural modulus of a specimen generally increases with increasing strain rate (S. Liu et al., 2018).

4.2.3 Compressive Properties (ASTM D 645, ISO 75-1 AND 75-2)

Compression properties describe the behavior of a material to withstand axial loads without failure. The compression is applied at a relatively low and uniform rate of loading as per ASTM test standards. Recently, a study on polylactic acid (PLA) specimen manufactured by fused deposition modelling found that the compression strength is closely aligned with ASTM D 645 standards, approving the viability of PLA in structural applications (Shivraj Narayan et al., 2024). However, the compressive strength of plastics has limited design value. The compressive strength and modulus values are obtained from the stress–strain curve in a similar way as in the case of tensile and flexural testing. Molecular orientation could affect the compressive strength of a material; for example, the carbon/carbon composites reinforced with CNTs exhibit enhanced out-of-plane compressive strength, influenced by the orientation and length of the nanotubes (Yu et al., 2014).

4.2.4 Creep Properties

The phenomenon of deformation of a material under load with time is called creep. It is governed by the constant load that continues to deform slowly with time indefinitely or until rupture or yielding causes failure. All plastic materials creep to a certain extent, and the degree of creep (deformation) depends upon the type of plastic, amount of load, time, and temperature. Under standard testing conditions (basically stress and temperature), all plastics have a characteristic type of creep behavior, as shown in Figure 4.10.

The creep properties of PNC are critical for their performance in various applications and are influenced by factors such as nanofiller type, content, and processing conditions. The creep values are obtained by applying a constant load to the test specimen in tension, compression, or flexure and measuring the deformation as a function of time. The incorporation of nanofillers significantly enhances the creep resistance of polymers. Kordzangeneh et al. (2024) reported that on addition of 0.3 wt% multi-walled CNTs to epoxy resin resulted in a 24.6% reduction in creep strain and a 74.1% decrease in steady-state creep rate compared to neat epoxy (Kordzangeneh et al., 2024). This enhancement is attributed to the improved mechanical properties and reduced viscoelastic behavior. Similarly, polyethylene (PE) nanocomposites incorporation of wt% clay improved creep life, while higher concentrations led to decreased creep strength, especially at elevated temperatures (Valojerdi et al., 2023). Creep behavior is generally temperature-dependent, and higher temperatures decrease creep strength and accelerate plastic deformation (Valojerdi et al., 2023). For instance, PE nanocomposites exhibited significant changes in creep behavior at 50°C compared to 25°C. Recently, advanced computational models have been designed to determine the creep behavior of 3D nanocomposites. Jayswal et al. (2023) developed one such

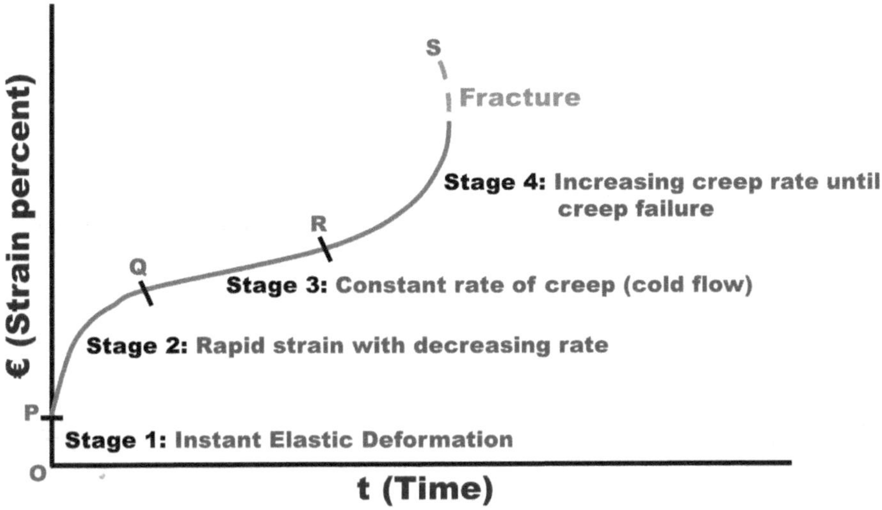

FIGURE 4.10 Standard creep curve.

model called the Generalized Voigt-Kelvin model that shows good agreement with experimental results (Jayswal et al., 2023). This modelling aids in understanding and optimizing creep properties.

4.2.5 Impact Properties

The impact property of a material is directly related to the overall toughness of the material. Toughness is defined as the ability of a polymer to absorb applied load. The area under the stress–strain curve is directly proportional to the toughness of a material. Impact energy is a measure of toughness. The higher the impact energy of a material, higher the toughness and vice versa. Impact resistance is the ability of a material to resist breaking under a shock loading or the ability to resist fracture under stress applied at high speed. The molecular flexibility plays a vital role in determining the relative brittleness or toughness of the material. For example, in polystyrene and acrylics, the molecular segments are unable to disentangle and respond to the rapid application of mechanical stress, and the impact produces brittle failure. However, flexible polymers like plasticized vinyls have high-impact behavior due to the ability of the large segments of molecules to disentangle and respond rapidly to mechanical stress.

The impact properties of a polymer can be modified simply by adding filler or impact modifiers such as butadiene rubber or certain acrylic polymers. The addition of plasticizers also improves the impact behavior at the cost of rigidity. The use of fibrous fillers can also aid in enhancing the impact energy as they act as stress transfer agents. Figure 4.11 shows different types of impact testing methods, such as Izod-Charpy Impact test (ASTM D-256, ASTM D 4812, ISO 174), Chip Impact test (ASTM D 4508), and Tensile Impact test (ASTM D 1822). The impact properties of a polymer are affected by several factors such as the rate of loading, notch sensitivity,

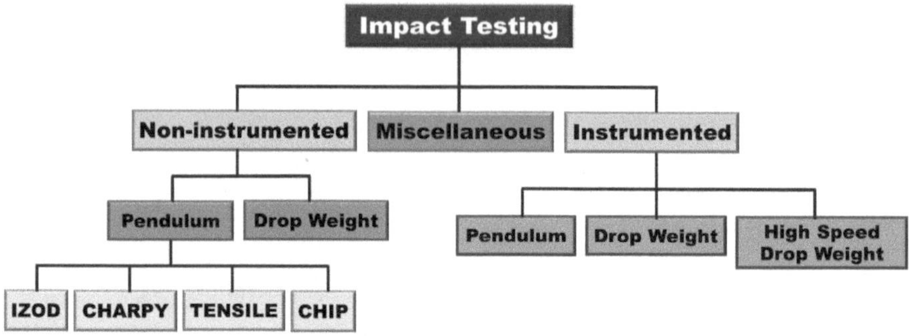

FIGURE 4.11 Types of impact test.

orientation, processing conditions and type, degree of crystallinity, molecular weight, and methods of loading.

Speed of loading: At low speed of loading even the brittle material can still have good impact strength while at high speed of loading even the rubber-like materials could show impact failure. Thus, the speed of loading plays a key role in impact testing.

Temperature: Like other mechanical properties, the impact behavior of a material is also affected by temperature. At lower temperatures, the impact resistance is reduced drastically.

Orientation: The orientation of polymer molecules has a major effect on the impact behavior of the polymer. Molecular orientation in the form of films and fibers may provide extra strength and toughness over isotropic material. The impact resistance is observed to be always higher in the direction of flow. However, such directional orientation of polymer molecules can be very fatal in a molded part since the impact stresses are usually multiaxial.

Notch sensitivity: A notch in a test specimen or a sharp corner in a fabricated part drastically lowers the impact energy. A notch creates a localized stress concentration, and hence, the part fails under impact loading. All plastics are notch sensitive, and the sensitivity varies with the type of plastic. Both notch depth and notch radius affect the impact behavior of materials. For example, a large radius of curvature at the base of the notch will have a lower stress concentration and therefore a higher impact energy of the base material.

Degree of crystallinity and molecular weight: With the increase in crystallinity, the impact resistance decreases, which increases the probability of brittle failure. A reduction in the average molecular weight tends to reduce the impact behavior and vice versa.

4.2.6 Shear Strength (ASTM D 732)

Shear strength of a plastic material is defined as the ability to withstand the maximum load required to shear the specimen so that the moving portion completely clears the stationary portion. The shear strength test is carried out by forcing a standardized punch at a specified rate through a sheet of plastic until the two portions of the specimen completely separate. The shear strength is determined by dividing the force required to shear the specimen by the area of the sheared edge and is expressed as lb/in^2.

Shear strength data is of great importance to a designer of film and sheet products that tend to be subjected to such shear loads. Most large molded and extruded products are usually not subjected to shear loads. For design purposes, shear strength is generally considered to be essentially equal to one-half the tensile strength.

4.2.7 Abrasion Properties

Abrasion resistance is defined as the ability of a polymeric material to resist abrasion and is measured by its loss in weight when abraded with an abraser. The most widely accepted abraser in the industry is called the Taber abraser. A variety of wheels with varied degrees of abrasiveness is available. For soft materials, less abrasive wheels with a smaller load on the wheels may be used. The test specimen is usually a 4 in. diameter disc or a 4 in.^{2} plate having both surfaces substantially plane and parallel. A ½ in. diameter hole is drilled in the center. Specimens are conditioned employing standard conditioning practices before testing. The grade of the abrasive wheel and the amount of load at which the test is carried out is always reported along with the results. Test methods such as ASTM D 1044 are developed for estimating the resistance of transparent plastic materials to one kind of abrasion by measurement of its optical effects. Another method designated as ASTM D 1242 is used to study the resistance of plastic material to abrasion by measuring the volume loss when a flat specimen is subjected to abrasion with loose abrasive or bonded abrasive on cloth or paper.

4.2.8 Fatigue Resistance

Fatigue resistance and fracture toughness are vital for assessing the durability and long-term performance of nanocomposites under cyclic loading and crack propagation conditions. Fatigue testing involves subjecting a specimen to repeated loading cycles until failure, providing insights into the material's endurance limit and fatigue life (ASTM D7741-12). Fracture toughness, often measured using the critical stress intensity factor, indicates the resistance of a material to crack propagation (ASTM E344-12). Nanocomposites reinforced with graphene or CNTs often exhibit enhanced fatigue and fracture toughness due to the effective stress transfer and energy dissipation mechanisms provided by the nanofillers (Khalid et al., 2023). For example, graphene oxide-reinforced PNCs have demonstrated significant improvements in fracture toughness, attributed to toughening mechanisms such as crack deflection and bridging.

4.2.9 Hardness Tests

Hardness is defined as the resistance of a material to deformation, particularly permanent deformation, indentation, or scratching. Hardness is purely a relative term and should not be confused with the wear and abrasion resistance of plastic materials. For example, polystyrene has a high Rockwell hardness value but poor abrasion resistance. Hardness tests can differentiate the relative hardness of different grades of a particular plastic. However, it is not valid to compare the hardness of various plastics entirely based on one type of test, since elastic recovery along with hardness is involved. Since plastic materials vary considerably concerning hardness, one type of hardness test does not cover the entire range of hardness properties encountered. Two of the most commonly used tests for plastic are the Rockwell hardness tests (ASTM D 785) and the Durometer hardness tests (ASTM D 2240, ISO 868). The former is used for relatively hard plastics such as acetals, nylons, acrylics and polystyrene, while the latter is used for softer materials like flexible PVC, thermoplastic rubbers, and PE. Another hardness test is devised mainly to measure the hardness of both reinforced and nonreinforced rigid plastics, known as Barcol Hardness (ASTM D 2583).

The factors that affect the hardness test of a material include temperature and humidity, surface conditions of the specimen, filler, and anisotropy.

Temperature and humidity: The hardness of all plastic materials is affected by the change in temperature and humidity. Specimens tested at higher than specified temperatures tend to indicate a lower hardness value.

Surface conditioning of specimen: The surface finish has a significant effect on hardness test results. A smooth molded surface yields a higher value than a machined surface.

Filler: The hardness value may vary depending on the type and amount of filler present in the resin. The variation in hardness reading may be caused by the difference in hardness between resin and filler material.

Anisotropy: Plastic materials with anisotropic characteristics may cause indentation hardness to vary with the direction of testing.

Shear relaxation: It is defined as a gradual decrease in stress with time under constant deformation. This characteristic is measured by applying a fixed amount of strain (deformation) to a specimen and measuring the load required to maintain it as a function of time. This is in contrast to creep measurement, where a fixed amount of load is applied and the resulting deformation is measured as a function of time.

4.3 TYPES OF FABRICATION TECHNIQUES

The fabrication of PNC employs various techniques to enhance its mechanical properties through nanoscale reinforcements. These methods are crucial for optimizing the overall performance of materials for diverse applications.

4.3.1 Solution Casting

This method involves dissolving polymers and incorporating nanofillers together into the solvent in which the polymeric material is easily soluble or blends, called in-situ synthesis of PNC. This method allows for a uniform distribution of the nanofillers within the polymer matrix, which is crucial for enhancing the mechanical properties of the resulting nanocomposites. Champa-Bujaico et al. (2024) fabricated multiscale poly(3-hydroxybutyrate) (P3HB)-based nanocomposite reinforced with MWCNTs, WS2 nanosheets and sepiolite (SEP) nanoclay. This resulted in significant improvements in the stiffness and ductility of the fabricated nanocomposite (Champa-Bujaico et al., 2024). The solution casting method facilitated a synergistic reinforcement effect among the different nanofillers. Such enhancements are attributed to the effective integration of the nanofillers achieved through the solution casting process. Therefore, the solution casting method enables homogeneous dispersion of nanofillers and leads to significant improvements in mechanical properties.

4.3.2 Extrusion

Extrusion is a manufacturing process where materials are pushed through a die to create continuous shapes. In the context of PNC, it involves mixing polymers with nanofillers and then forming them into desired shapes. It is a widely used technique that allows for the continuous production of nanocomposites, enhancing their mechanical and thermal properties through uniform dispersion of nanofillers. It enables the thorough mixing of nanofillers with a polymeric matrix, which is crucial for uniform dispersion to obtain nanocomposites with superior mechanical properties. During extrusion, the temperature and shear forces can be controlled, which affects the morphology of nanocomposites. This control is vital as it facilitates the optimization, leading to better interfacial adhesion between polymer and nanofillers, resulting in enhanced material properties (such as mechanical, thermal, and barrier properties; Zachariah et al., 2024).

Extrusion is the best technique for the bulk production of nanocomposites while maintaining quality. It is a scalable process, thus making it suitable for industrial applications. It is also stated that extrusion is cost-effective in comparison to other fabrication methods. The cost-effectiveness of the extrusion process, combined with the enhancement of the mechanical properties of nanocomposites, makes it an attractive tool for producing high-performance nanocomposites for diverse applications.

4.3.3 Layer-by-Layer Assembly

The layer-by-layer (L-b-L) fabrication technique is a significant method used in the development of PNCs, particularly for enhancing their mechanical properties. This technique enables precise control over the thickness and composition of nanocomposite layers, improving barrier and mechanical properties (Zachariah et al., 2024). Using this method, it is possible to create a composite with tailored properties by alternating layers of different materials, which can lead to improved mechanical strength and flexibility. The layering process enhances interfacial interactions between the different materials used in the nanocomposite. Stronger interfacial bonding can lead to better load transfer between layers, which is crucial for enhancing mechanical properties such as tensile strength and toughness.

The L-b-L technique increases the surface area of the nanocomposite, which can improve the overall mechanical performance. A larger surface area allows for more effective stress distribution throughout the material, reducing the likelihood of failure under mechanical stress. While the focus is on mechanical properties, the L-b-L technique also contributes to improved barrier properties, which are essential for food packaging applications. Enhanced barrier properties can indirectly support mechanical integrity by preventing degradation of the material over time (Zachariah et al., 2024).

While these techniques significantly enhance mechanical properties, challenges remain in achieving uniform dispersion and scalability, which can limit the practical applications of PNCs in industry and have been discussed later in this chapter.

4.4 FACTORS INFLUENCING THE ENHANCEMENT OF MECHANICAL PROPERTIES

The enhancement of mechanical properties in PNCs is governed by several key factors that are crucial for optimizing the performance of these advanced materials. These factors are discussed below:

Nanofiller type and size: Nano-sized fillers such as nanoparticles, nanotubes, and nanofibers generally provide better reinforcement than micro-sized fillers. Their small size ranges from 1 to 100 nm, and they have a large surface area and high aspect ratio, improving their dispersion and interfacial bonding with the polymeric matrix. Homogeneous dispersion of nanofillers ensures uniform distribution of stress throughout the polymer matrix, resulting in better load transfer (Šupová et al., 2011). Different nanoparticles, such as graphene, CNTs and silica, aid in yielding superior mechanical properties. For example, the incorporation of multiwalled CNTs (MWCNTs) into polyurethane led to a 21% increase in ultimate tensile strength due to enhanced dispersion and bonding (Gohar et al., 2023). Kontou et al. (2023) have studied that nano-sized silica particles provide higher reinforcement in PNCs, yielding enhanced mechanical enhancements (Kontou et al., 2023). However, further studies related to the combined effect of matrix type, nanoparticle size, and dispersion quality still need attention.

Nanofiller dispersion: These factors play a critical role in the enhancement of mechanical properties. Uniform distribution of nanofillers within the polymeric matrix can be quite challenging at times. Improper dispersion could lead to deformations, voids, and aggregation of nanofillers. This significantly affects the nanocomposite properties and their performance. Poorly dispersed nanofillers lead to weak interfacial bonding between the matrix and the nanofillers, causing lower tensile strength, stiffness and impact resistance. Several characteristics of nanocomposites rely on uniform filler distribution, such as thermal and electrical conductivity. This is due to the creation of hotspots with poor conductivity, leading to inefficient heat/ion transfer. Additionally, uneven dispersion could also create pathways for gas or moisture permeation, compromising the effectiveness of the barrier. Aggregation of nanofillers can increase the viscosity of the composite material, making processing more difficult and potentially leading to defects during manufacturing. Tiwari and Panda (2024) in their work had shown an increase in the stiffness of PNCs when incorporated with well-aligned graphene nanoplatelets (GNPs) and Fe_3O_4-GNP. It was also observed that there was a significant 31% increase in Young's modulus of the nanocomposite due to aligned Fe_3O_4-GNP (Tiwari & Panda, 2024). In another study, E glass fiber and nanocarbon fibers were reinforced into the epoxy matrix to fabricate PNC using mechanical compression molding and techniques like

L6 Taguchi, and design of experiments (DOEs) was used to systematically optimize the processing conditions. These factors enabled optimization of processing conditions as well as the enhancement of mechanical properties of the resultant nanocomposite (Vinayaka et al., 2023).

Interfacial bonding: Grafting is a most prominent technique that is used to enhance the physical, chemical, and morphological properties of the polymeric materials (Purohit et al., 2023). It is a technique in which monomers are covalently bonded and polymerized as side chains onto the main polymer chain (Tauqir A. Sherazi, 2016). Similarly, direct grafting of nanofillers onto a polymer matrix enables better load transfer between the matrix and filler, along with improved dispersion of nanofillers. A study conducted by Ries et al. (2023) discusses a promising increase in the strength of the PNCs compared to those with untreated fillers, demonstrating that effective interfacial bonding via grafting plays a vital role in mechanical properties enhancement (Ries et al., 2023). Interfacial interactions, principally dispersion interactions, enhance heterogeneous nucleation by increasing glycine concentration at nonpolar surfaces, facilitating kinetic and structural effects for nucleation enhancement (McKechnie et al., 2022). In another case, enhanced interfacial H-bond networks promote glycan–glycan enrichment efficiency and enable detailed glycosylation profiling in biological systems (J. Liu et al., 2023).

4.5 ENHANCEMENT OF MECHANICAL PROPERTIES

4.5.1 Reinforcement of Nanofillers

It is the most common and widely exploited method that effectively combines the properties of nanofillers and polymeric materials to generate nanocomposites with enhanced mechanical strength. This enhancement is attributed to the unique characteristics of nanofillers, including their high surface area and specific geometrical properties, which facilitate better dispersion and interaction within the polymer matrix. The homogeneous dispersion of nanofillers within the polymeric matrix ensures that the stress is evenly distributed throughout the material, leading to improved load transfer from the polymer matrix to the nanofillers.

Techniques such as solution blending, melt mixing, and in-situ polymerization are commonly used to achieve optimal dispersion. The geometry of nanofillers, such as CNTs and graphene, plays a crucial role in determining the mechanical properties of composites. For instance, the Young's modulus is influenced more by the aspect ratio and orientation of the nanofillers than their intrinsic mechanical properties (Li et al., 2024). Hybrid systems combining different types of nanofillers can create synergistic effects, significantly enhancing properties like tensile strength and elongation at break (Afsharpour & Kazemi, 2024).

4.5.1.1 Types of Nanofillers

Carbon-based nanofillers: Carbon-based nanofillers, such as graphene and CNTs, are widely studied for their remarkable mechanical properties. Graphene, with its high specific surface area and intrinsic strength, can significantly enhance the tensile strength and modulus of PNCs when well-dispersed and properly aligned within the matrix (Lee et al., 2022; Tarhini & Tehrani-Bagha, 2019). Similarly, CNTs, known for their high aspect ratio and extraordinary mechanical properties, provide excellent reinforcement to polymer matrices, improving tensile strength, modulus, and toughness (Arash et al., 2014; Nurazzi et al., 2021).

Metal oxide nanoparticles: Metal oxide nanoparticles, such as silica, alumina, and titanium dioxide, are also used to enhance the mechanical properties of PNCs. These nanoparticles can improve the stiffness, hardness, and impact resistance of the composites. For instance, silica nanoparticles have been shown to enhance the tensile strength and modulus of PNCs due to their high surface area and ability to form strong interfacial bonds with the polymer matrix (Shi et al., 2019)

Clay and silicate nanofillers: Clay and silicate nanofillers, including montmorillonite and layered silicates, are effective in enhancing the mechanical properties of PNCs. These nanofillers provide significant improvements in tensile strength, modulus, and barrier properties due to their high aspect ratio and ability to exfoliate within the polymer matrix (Mittal, 2009). Exfoliated clay nanocomposites exhibit enhanced stiffness and strength, attributed to the strong interfacial interactions and effective stress transfer between the polymer matrix and the nanofillers (Baniasadi et al., 2021; Franco-Urquiza, 2021; Mittal, 2009).

Other emerging nanofillers: Other emerging nanofillers, such as CNCs, boron nitride nanosheets, and MXenes, are gaining attention for their potential to enhance the mechanical properties of PNCs. CNCs, derived from renewable sources, offer high strength and stiffness, making them attractive for sustainable nanocomposites (Norizan et al., 2022). Boron nitride nanosheets provide excellent thermal conductivity and mechanical reinforcement, while MXenes, a family of two-dimensional transition metal carbides and nitrides, offer a unique combination of electrical conductivity and mechanical strength (Protyai & Bin Rashid, 2024; C. (John) Zhang et al., 2020).

Surface functionalization: The interfacial interaction between the nanofillers and the polymer matrix is another crucial factor influencing the mechanical properties of nanocomposites. Strong interfacial bonding facilitates efficient stress transfer from the polymer matrix to the nanofillers, enhancing the overall mechanical performance. Methods to improve interfacial interactions include surface functionalization of nanofillers, the use of coupling agents, and the introduction of compatibilizers (Goh et al., 2020). For instance, silane coupling agents are often used to enhance the adhesion between silica nanoparticles and polymer matrices, leading to significant improvements in tensile strength and modulus (Chuang et al., 2018; Xie et al., 2010).

4.6 APPLICATIONS OF PNC

4.6.1 Automobile

Nanocomposites are increasingly used in the automotive industry due to their superior mechanical properties, including high strength, stiffness, and impact resistance. These materials help in reducing vehicle weight, which in turn enhances fuel efficiency and reduces emissions. For instance, CNT-reinforced PNCs are used in automotive components such as bumpers, dashboards, and interior panels, providing improved crashworthiness and durability (Khan et al., 2024). Additionally, nanocomposites with graphene have shown potential in enhancing the thermal and mechanical properties of tires, improving their performance and lifespan (Sayfo et al., 2023).

4.6.2 Packaging

In the packaging industry, nanocomposites are valued for their enhanced mechanical properties and barrier performance. The incorporation of nanoclays into polymer matrices improves the tensile strength, modulus, and impact resistance of packaging materials while also providing excellent barrier properties against gases and moisture (Franco-Urquiza, 2021). For example, PE nanocomposites reinforced with montmorillonite clay are used in food packaging to extend shelf life and maintain product quality (Nath et al., 2022). These materials offer lightweight, durable, and recyclable packaging solutions.

4.6.3 Agricultural

Nanocomposites find applications in agriculture due to their enhanced mechanical properties, which contribute to the durability and functionality of agricultural films, nets, and containers. Silica and clay nanoparticle-reinforced PNCs are used in greenhouse films, providing improved strength and resistance to environmental degradation (Chuang et al., 2018; Nath et al., 2022). These films help in regulating temperature and light transmission, promoting better crop growth. Additionally, nanocomposite-based drip irrigation pipes offer superior mechanical strength and flexibility, enhancing their longevity and performance.

4.6.4 Construction

In the construction industry, nanocomposites are utilized for their enhanced mechanical properties, such as increased strength, stiffness, and impact resistance. Nanoparticles like silica, alumina, and CNTs are incorporated into cementitious materials to improve their mechanical performance and durability (Mishra, 2022). For instance, the addition of silica nanoparticles to concrete can enhance its compressive strength and reduce permeability, leading to more durable and resilient structures (Arif et al., 2023). Nanocomposite coatings are also used to protect building materials from corrosion and wear.

4.6.5 Defense

The defense sector benefits from the superior mechanical properties of nanocomposites, which provide lightweight and high-strength materials for various applications. CNT-reinforced composites are used in body armor and helmets, offering enhanced ballistic protection and reduced weight compared to traditional materials (Bhinder et al., 2021). Nanocomposites are also used in military vehicle components and protective coatings, providing improved mechanical performance, durability, and resistance to extreme conditions.

4.6.6 Automotive and Aerospace

In the automotive and aerospace industries, nanocomposites are crucial for developing lightweight and high-performance materials. The incorporation of nanofillers into polymer matrices significantly enhances the mechanical properties, including tensile strength, modulus, and impact resistance. These properties are essential for structural components, engine parts, and interior materials in both automobiles and aircraft, contributing to fuel efficiency, safety, and overall performance. For example, CNT-reinforced composites are used in aircraft fuselage and wing structures, providing improved strength-to-weight ratios and resistance to fatigue and fracture (Parveez et al., 2022).

4.6.7 Biomedical Devices

The biomedical field leverages nanocomposites for their excellent mechanical properties and biocompatibility. Nanocomposites reinforced with nanoparticles such as hydroxyapatite, silica, and CNTs are used in medical implants, prosthetics, and dental materials, offering improved mechanical strength and durability. For instance, hydroxyapatite-reinforced PNCs are used in bone grafts and dental restorations, providing enhanced mechanical properties and promoting osteointegration (Abbas et al., 2023). Additionally, nanocomposite-based scaffolds are employed in tissue engineering, offering suitable mechanical properties and promoting cell growth and tissue regeneration (Wahid et al., 2018).

4.6.8 Electronics and Sensors

In the electronics and sensors industry, nanocomposites are utilized for their enhanced mechanical properties and multifunctionality. Conductive nanofillers like graphene and CNTs improve the mechanical strength, flexibility, and electrical conductivity of PNCs, making them suitable for flexible electronics, wearable devices, and sensors (Schintke et al., 2023; Zhao et al., 2023). For example, graphene-reinforced nanocomposites are used in flexible displays and touchscreens, providing excellent mechanical strength and flexibility (Schintke et al., 2023). Additionally, nanocomposite-based sensors offer high sensitivity and durability for various applications, including environmental monitoring and biomedical diagnostics.

In summary, the applications of nanocomposites span across various industries, leveraging their enhanced mechanical properties to develop high-performance, durable, and multifunctional materials. The incorporation of different types of nanofillers plays a crucial role in tailoring the mechanical properties of nanocomposites for specific applications (Table 4.2).

4.7 CONCLUSION: CHALLENGES AND FUTURE PERSPECTIVES

Enhancement in the mechanical properties of nanocomposites is promising; however, some challenges require critical attention for their future advancements, which are as discussed below.

TABLE 4.2
Summary of Mechanical Property Enhancement for Different Polymer Nanocomposites

Polymer	Nanofiller	Method of Fabrication	Properties Enhanced	Applications	References
Polypropylene (PP)	Multiwall carbon nanotubes (CNTs)	Melt compounding process, hot pressing and solid-state drawing	Tensile strength, elastic modulus, impact strength	Automotive parts, electrical components	Deng et al. (2010)
Polyvinyl alcohol (PVA)	Graphene oxide (GO)	Solution blending, compression molding	Young's modulus, tensile strength	Packaging, construction materials	X. Zhang et al. (2019)
Bio-based polyamide (PAs)	Nanoclay	Extrusion molding	Toughness, barrier properties	Packaging	Battegazzore et al. (2019)
Polyvinyl alcohol hydrogel (PVA)	Silica nanoparticles, urea	Freeze-thaw and soaking process	Tensile strength, toughness	Biomedical application (Cartilage recovery)	Tan et al. (2024)
Polystyrene (PS)	Carbon nanowires (CNWs)	Solution precipitation method	Thermal and tensile properties	Electronics	Kostas et al. (2020)

4.7.1 Agglomeration and Dispersion of Nanofillers

Achieving uniform dispersion of nanofillers within the polymer matrix is one of the primary challenges. It has been observed that nanofillers tend to agglomerate due to van der Waals forces and other intermolecular interactions, causing nonuniform dispersion and unwanted agglomerations. This creates stress concentration points that weaken the overall mechanical characteristics of the nanocomposite (Wang et al., 2011). This makes it difficult to achieve homogeneous dispersion of nanofillers inside the polymer matrix. Improper and poor dispersion of nanofillers affects both directional (such as tensile strength and conductivity) and nondirectional (like hardness or elasticity) properties of the PNC. Directional properties are governed by the unique structures and compositions of the nanofillers and polymer matrix. These are mainly influenced by anisotropic fillers such as CNT (CNTs), while nondirectional properties are governed by isotropic fillers such as carbon black (Agnelli et al., 2018). Aggregation of nanofillers leads to irreversible deformation, resulting in low strength, creation of interfacial voids and poor gas or moisture barrier properties. Techniques such as surface functionalization, ultrasonication, and high-shear mixing of nanofillers can be employed to eliminate these issues. However, large-scale dispersion still remains a challenge for achieving consistency. Therefore, good interfacial compatibility between the nanofillers and the polymer matrix is vital for the fabrication of nanocomposites.

4.7.2 Interfacial Compatibility

Weak interfacial interaction among the nanofillers and matrix could lead to poor load transfer and reduced mechanical properties. Surface functionalization of nanofillers via suitable chemical methods is an effective way of increasing chemical compatibility between the two (nanofiller and polymer matrix). This would lead to better interfacial interactions between both phases and better enhancement of the characteristic properties of PNC. However, the suitability of functionalization could vary depending on the type of polymer matrix and nanofiller used. Additionally, the development of hybrid nanocomposite systems, such as combining more than two materials, could also facilitate additional enhancement of mechanical properties and performance of nanocomposites. Also, understanding the mechanism behind the mechanical properties of nanocomposites can help in designing more effective techniques for their fabrication.

4.7.3 Environmental and Health Impacts

It is another primary and crucial considerations that need to be addressed, especially when microplastic waste generation and management have been a major concern globally (Rafiq & Xu, 2023). Therefore, it is significant to assess the toxicity, biocompatibility, and environmental impact of nanofillers and PNC to ensure their safe use. During the synthesis of nanocomposites, monitoring of environmental and sustainable factors becomes of utmost importance to minimize potential risks of ecological harm and health risks to humans. Thus, future research should focus on developing environmentally friendly and biodegradable nanocomposites.

4.7.4 Scale-Up Process

To scale up the production of PNC while maintaining its enhanced mechanical properties is significantly challenging. The development of scalable and cost-effective manufacturing processes that could address above-mentioned challenges is very essential for widespread adoption of PNC in diverse industries. This is because laboratory-scale fabrication techniques often do not translate well to industrial production due to several issues like inconsistent quality, high costs, and complex processing requirements. Hence, collaborative efforts between academics, industries, and government agencies are necessary to address these challenges and facilitate the commercialization of high-performance nanocomposites.

In conclusion, although substantial progress has been made in the enhancement of mechanical properties of the nanocomposites, several challenges remain. Addressing these significant challenges becomes essential to achieve commercialization and widespread adoption of PNC without compromising their enhanced mechanical characteristics. Additionally, the development of advanced synthesis techniques is pivotal for overcoming these challenges by reducing environmental impact and maintaining sustainable practices. This would aid in discovering the full potential of PNC in various applications. Thus, enhancement of the mechanical properties of PNC holds a promising future advancement in this field.

ACKNOWLEDGMENT

The authors are thankful to the Netaji Subhas University of Technology for maintaining academic and research integrity.

REFERENCES

Abbas M, Alqahtani MS, and Alhifzi R (2023) Recent developments in polymer nanocomposites for bone regeneration. *Int J Mol Sci*, 24(4): 3312.

Afsharpour M, and Kazemi B (2024) Nanofillers in additives for rubbers. In: Mallakpour S, Hussain CM (eds.) *Handbook of Nanofillers*. Springer, Singapore, pp. 1–38.

Agnelli S, Pandini S, Torricelli F, Romele P, Serafini A, Barbera V, and Galimberti M (2018) Anisotropic properties of elastomeric nanocomposites based on natural rubber and sp2 carbon allotropes. *Exp Polym Lett*, 12(8): 713–730.

Al-Hakimi A, Asnag G, Alminderej F, Alhagri I, Al-Hazmy S, and Qahtan T (2023) Enhancing the structural, optical, thermal, and electrical properties of PVA filled with mixed nanoparticles (TiO2/Cu). *Crystals*, 13(1): 135.

Arash B, Wang Q, and Varadan VK (2014) Mechanical properties of carbon nanotube/polymer composites. *Sci Rep*, 4(1): 6479.

Arif M, Hasan SD, and Siddiqui S (2023) Effect of nano silica on strength and permeability of concrete. *Mater Today Proc*, 2214–7853.

Bal S, and Samal SS (2007) Carbon nanotube reinforced polymer composites—a state of the art. *Bull Mater Sci*, 30(4): 379–386.

Baniasadi H, Trifol J, Ranta A, and Seppälä J (2021) Exfoliated clay nanocomposites of renewable long-chain aliphatic polyamide through in-situ polymerization. *Compos B Eng*, 211: 108655.

Battegazzore D, Sattin A, Maspoch ML, and Frache A (2019) Mechanical and barrier properties enhancement in film extruded bio-polyamides with modified nanoclay. *Polym Compos*, 40(7): 2617–2628.

Bhinder J, Verma SK, and Agnihotri PK (2021) Qualifying carbon nanotube reinforced polyurethane foam as helmet inner liner through in-situ, static and low velocity impact testing. *Mater Sci Eng B*, 274: 115496.

Chadha U, Selvaraj SK, Ashokan H, Hariharan SP, Mathew PV, Venkatarangan V, and Paramasivam V (2022) Complex nanomaterials in catalysis for chemically significant applications: from synthesis and hydrocarbon processing to renewable energy applications. *Adv Mater Sci Eng*, 2022: 1–72.

Champa-Bujaico E, Díez-Pascual AM, Lomas RA, and Garcia-Diaz P (2024) Optimization of mechanical properties of multiscale hybrid polymer nanocomposites: a combination of experimental and machine learning techniques. *Compos B Eng*, 269: 111099.

Chuang W, Geng-sheng J, Lei P, Bao-lin Z, Ke-zhi L, and Jun-long W (2018) Influences of surface modification of nano-silica by silane coupling agents on the thermal and frictional properties of cyanate ester resin. *Results Phys*, 9: 886–896.

Darwish MSA, Mostafa MH, and Al-Harbi LM (2022) Polymeric nanocomposites for environmental and industrial applications. *Int J Mol Sci*, 23(3): 1023.

Deng H, Bilotti E, Zhang R, and Peijs T (2010) Effective reinforcement of carbon nanotubes in polypropylene matrices. *J Appl Polym Sci*, 118(1): 30–41.

Franco-Urquiza EA (2021) Clay-based polymer nanocomposites: essential work of fracture. *Polymers*, 13(15): 2399.

Goh PS, Wong KC, Yogarathinam LT, Ismail AF, Abdullah MS, and Ng BC (2020) Surface modifications of nanofillers for carbon dioxide separation nanocomposite membrane. Symmetry, 12(7): 1102.

Gohar GA, Akhtar A, Raza H, Mustafa G, Fatima M, Rehman HU, Aslam MW, Haq A ul, and Manzoor W (2023) Achieving enhanced tensile properties of polyurethane-multiwall carbon nanotubes nanocomposites. *Nano Express*, 4(4): 045013.

Guan LZ, and Tang LC (2021) Dispersion and alignment of carbon nanotubes in polymer matrix. In: Abraham J, Thomas S, Kalarikkal N (eds.) *Handbook of Carbon Nanotubes*. Springer, Cham, pp. 1–35.

Jayswal A, Liu J, Harris G, Mailen R, and Adanur S (2023) Creep behavior of 3D printed polymer composites. *Polym Eng Sci*, 63(11): 3809–3818.

Kamal A, Ashmawy MSS, Algazzar AM, and Elsheikh AH (2022) Fabrication techniques of polymeric nanocomposites: a comprehensive review. *Proc Inst Mech Eng Pt C J Mech Eng Sci*, 236(9): 4843–4861.

Khalid MY, Kamal A, Otabil A, Mamoun O, and Liao K (2023) Graphene/epoxy nanocomposites for improved fracture toughness: a focused review on toughening mechanism. *Chem Eng J Adv*, 16: 100537.

Khan F, Hossain N, Mim JJ, Rahman SM, Iqbal MJ, Billah M, and Chowdhury MA (2024) Advances of composite materials in automobile applications – a review. *J Eng Res*, 2307–1877.

Kontou E, Christopoulos A, Koralli P, and Mouzakis DE (2023) The effect of silica particle size on the mechanical enhancement of polymer nanocomposites. *Nanomater*, 13(6): 1095.

Kordzangeneh D, Khoramishad H, and Fatolahi AR (2024) Post-creep residual tensile properties of multi-walled carbon nanotube/epoxy nanocomposites. *Proc Inst Mech Eng Pt L J Mater Des* Appl, 238(11): 2131–2141.

Kostas V, Baikousi M, Barkoula NM, Giannakas A, Kouloumpis A, Avgeropoulos A, Gournis D, and Karakassides MA (2020) Synthesis, characterization and mechanical properties of nanocomposites based on novel carbon nanowires and polystyrene. *Appl Sci*, 10(17): 5737.

Lee SJ, Yoon SJ, and Jeon IY (2022) Graphene/polymer nanocomposites: preparation, mechanical properties, and application. *Polymers*, 14(21): 4733.

Li Z, Liu M, and Young RJ (2024) Dependence of the reinforcement of polymer-based nanocomposites upon the nanofiller geometry. *Nano Mater Sci,* doi.org/10.1016/j.nanoms.2024.04.014.

Liu J, Lv J, Liu Z, Fang Z, Lai C, Zhao S, Ye M, and Wang F (2023) Enhanced interfacial H-bond networks promote glycan–glycan recognition and interaction. *ACS Appl Mater Interfaces*, 15(14): 17592–17600.

Liu S, Zhu D, Yao Y, and Shi C (2018) Effects of strain rate and temperature on the flexural behavior of basalt and glass textile–reinforced concrete. *J Mater Civ Eng*, 30(8): 04018172.

Lu PJ, Huang SC, Chen YP, Chiueh LC, and Shih DYC (2015) Analysis of titanium dioxide and zinc oxide nanoparticles in cosmetics. *JFDA*, 23(3): 587–594.

Ma PC, Siddiqui NA, Marom G, and Kim JK (2010) Dispersion and functionalization of carbon nanotubes for polymer-based nanocomposites: a review. *Compos A: Appl Sci Manuf*, 41(10): 1345–1367.

McKechnie D, Mulheran PA, Sefcik J, and Johnston K (2022) Tuning interfacial concentration enhancement through dispersion interactions to facilitate heterogeneous nucleation. *JPC C*, 126(38): 16387–16400.

Meesorn W, Shirole A, Vanhecke D, de Espinosa LM, and Weder C (2017) A simple and versatile strategy to improve the mechanical properties of polymer nanocomposites with cellulose nanocrystals. *Macromolecules*, 50(6): 2364–2374.

Mishra G (2022) Co-effect of carbon nanotube and nano-sized silica on dispersion and mechanical performance in cementitious system. *DRM*, 127: 109162.

Mittal V (2009) Polymer Layered Silicate Nanocomposites: A Review. Materials, 2(3): 992–1057.

Müller K, Bugnicourt E, Latorre M, Jorda M, Echegoyen SY, Lagaron J, Miesbauer O, Bianchin A, Hankin S, Bölz U, Pérez G, Jesdinszki M, Lindner M, Scheuerer Z, Castelló S, and Schmid M (2017) Review on the processing and properties of polymer nanocomposites and nanocoatings and their applications in the packaging, automotive and solar energy fields. *Nanomater*, 7(4): 74.

Nan Z, Wei W, Lin Z, Chang J, and Hao Y (2023) Flexible nanocomposite conductors for electromagnetic interference shielding. *Nano Lett*, 15(1): 172.

Nath DRS, Pal K, and Sarkar P (2022) Nanoclay-based active food packaging systems: a review. *Food Packag Shelf Life*, 31: 100803.

Norizan MN, Shazleen SS, Alias AH, Sabaruddin FA, Asyraf MRM, Zainudin ES, Abdullah N, Samsudin MS, Kamarudin SH, and Norrahim MNF (2022) Nanocellulose-based nanocomposites for sustainable applications: a review. *Nanomaterials*, 12(19): 3483.

Nurazzi NM, Sabaruddin FA, Harussani MM, Kamarudin SH, Rayung M, Asyraf MRM, Aisyah HA, Norrahim MNF, Ilyas RA, Abdullah N, Zainudin ES, Sapuan SM, and Khalina A (2021) Mechanical performance and applications of CNTs reinforced polymer composites—a review. *Nanomaterials*, 11(9): 2186.

Parveez B, Kittur MI, Badruddin IA, Kamangar S, Hussien M, and Umarfarooq MA (2022) Scientific advancements in composite materials for aircraft applications: a review. *Polymers*, 14(22): 5007.

Protyai Md IH, and Rashid AB (2024) A comprehensive overview of recent progress in MXene-based polymer composites: their fabrication processes, advanced applications, and prospects. *Heliyon*, 10: e37030.

Purohit P, Bhatt A, Mittal RK, Abdellattif MH, and Farghaly TA (2023) Polymer Grafting and its chemical reactions. *Front Bioeng Biotechnol*, 10:1044927.

Rafiq A, and Xu JL (2023) Microplastics in waste management systems: a review of analytical methods, challenges and prospects. *Waste Manage*, 171: 54–70.

Ramakoti IS, Panda AK, and Gouda N (2023) A brief review on polymer nanocomposites: current trends and prospects. *J Polym Eng*, 43(8): 651–679.

Rashid AB, Haque M, Islam SMM, and Uddin Labib KMR (2024) Nanotechnology-enhanced fiber-reinforced polymer composites: Recent advancements on processing techniques and applications. *Heliyon*, 10(2): e24692.

Ries M, Reber S, Steinmann P, and Pfaller S (2023) Extending a generic and fast coarse-grained molecular dynamics model to examine the mechanical behavior of grafted polymer nanocomposites. *Forces Mech*, 12: 100207.

Saba N, Jawaid M, and Sultan MTH (2019) An overview of mechanical and physical testing of composite materials. In: *Mechanical and Physical Testing of Biocomposites, Fibre-Reinforced Composites and Hybrid Composites*, Mohammad Jawaid, Mohamed Thariq and Naheed Saba (Eds.). Elsevier, pp 1–12.

Saeed K, Khan I, Khan I, Ali N, Bilal M, and Akhter MS (2023) Graphene and carbon nanotubes-based polymer nanocomposites. In: *Smart Polymer Nanocomposites*, Nisar Ali, Muhammad Bilal, Adnan Khan, Ram K. Gupta, Tuan Anh Nguyen (Eds.). Elsevier, pp. 205–218.

Sayfo P, Pirityi DZ, and Pölöskei K (2023) Characterization of graphene-rubber nanocomposites: a review. *Mater Today Chem*, 29: 101397.

Schintke S, Mujovi B, Loiko D, and del Rossi S (2023) Graphite- and graphene-based polymer nanocomposites for flexible sensors and actuators in health care and soft robotics applications. *MSF*, 1107: 15–20.

Shahbaz A., Hussain N, Mahmood T, Iqbal HMN, Emran TB, Show PL, and Bilal M (2023) Polymer nanocomposites for biomedical applications. In *Smart Polymer Nanocomposites*, Nisar Ali, Muhammad Bilal, Adnan Khan, Ram K. Gupta, Tuan Anh Nguyen (Eds.), Elsevier, pp 379–394.

Shi X, Hassanzadeh-Aghdam MK, and Ansari R (2019) Viscoelastic analysis of silica nanoparticle-polymer nanocomposites. *Compos B Eng*, 158: 169–178.

Shivraj NY, Prakash JK, Narendra P, Kumar SA, and Barik K (2024) Compressive strength assessment in additively manufactured sustainable poly lactic acid specimens as per ASTM D695 standard. In: *16th International Conference on Materials Processing and Characterization (ICMPC 2024), E3S Web of Conferences, 552*, 01027.

Shrivastava A (2018) Additives for plastics. In: *Introduction to Plastics Engineering*. Elsevier, pp 111–141.

Singh K, Jaiswal R, Kumar R, Singh S, and Agarwal K (2023) Polymer-based nanocomposites as defence material. *Bull Mater Sci*, 46(2): 79.

Şomoghi R, Semenescu A, Pasăre V, Chivu OR, Niţoi DF, Marcu DF, and Florea B (2024) The impact of ZnO nanofillers on the mechanical and anti-corrosion performances of epoxy composites. *Polymers*, 16(14): 2054.

Song N., Gao Z, and Li X (2020) Tailoring nanocomposite interfaces with graphene to achieve high strength and toughness. *Sci Adv*, 6(42): eaba7016.

Šupová M, Martynková GS, and Barabaszová K (2011) Effect of nanofillers dispersion in polymer matrices: a review. *Sci Adv Mater*, 3(1): 1–25.

Tan X, Liang C, Bai S, Lan P, Ren Y, Zhao J, and Rong J (2024) PVA/silica hybrid hydrogel with ultra-high strength and toughness. *J Mater Sci*, 59(16): 6916–6928.

Tang Y, Zhao W, Zhu G, Tan Z, Huang L, Zhang P, Gao L, and Rui Y (2023) Nano-pesticides and fertilizers: solutions for global food security. *Nanomaterials*, 14(1): 90.

Tarhini AA, and Tehrani-Bagha AR (2019) Graphene-based polymer composite films with enhanced mechanical properties and ultra-high in-plane thermal conductivity. *Compos Sci Technol*, 184: 107797.

Tauqir AS (2016) Graft polymerization. In: Drioli E, Giorno L (eds.) *Encyclopedia of Membranes*, Vol. 8. Springer, Berlin, Heidelberg, pp. 886–887.

Tiwari A, and Panda SK (2024) Stiffness enhancement of polymer nanocomposites via graphene nanoplatelet orientation. *Polym Eng Sci*, 64(4): 1482–1493.

Tyagi M and Tyagi D (2014) Polymer nanocomposites and their applications in electronics industry. *Int J Electron Electr Eng*, 7: 603–608.

Valojerdi MA, Zebarjad SM, and Moghim MH (2023) A study on the role of clay nanoparticles and temperature on the creep behavior of polyethylene matrix nanocomposite. *J Mod Polym Chem Mater* 2:4.

Vaudagna MV, Aiassa V, Marcotti A, Pince Beti MF, Constantín MF, Pérez MF, Zoppi A, Becerra MC, and Silvero CMJ (2023) Titanium dioxide Nanoparticles in sunscreens and skin photo-damage. Development, synthesis and characterization of a novel biocompatible alternative based on their in vitro and in vivo study. *J Photochem Photobiol A*, 15: 100173.

Vinayaka N, Bodukuri AK, Jadhav GK, Padmamalini N, Pandey SK, Balasubramanian M, Immanuel Durai Raj J, Suresh Kumar M, and Singh B (2023) Analyze the mechanical characteristics of fabricated MMCs on nanocarbon influencing with polymer composites. *J Nanomaterials*, 2023: 1–10.

Wahid F, Khan T, Hussain Z, and Ullah H (2018) Nanocomposite scaffolds for tissue engineering; properties, preparation and applications. In: *Applications of Nanocomposite Materials in Drug Delivery*, Inamuddin, Abdullah M. Asiri and Ali Mohammad (Eds.). Elsevier, pp. 701–735.

Wang Y, Wu X, Yang W, Zhai Y, Xie B, and Yang M (2011) Aggregate of nanoparticles: rheological and mechanical properties. *Nanoscale Res Lett*, 6(1): 114.

Wing MY, Zhongzhen Y, Xaolin X, Qingxin Z, and Jun M (2003) Polymer nanocomposites and their applications. *HKIE Trans*, 10(4): 67–73.

Xie Y, Hill CAS, Xiao Z, Militz H, and Mai C (2010) Silane coupling agents used for natural fiber/polymer composites: a review. *Compos A: Appl Sci Manuf*, 41(7): 806–819.

Yu H, Lu J, Song Q, Li K, Li H, Fu Q, and Zhang L (2014) Compressive properties of carbon/carbon composites reinforced by carbon nanotubes with different orientations and lengths. *Vacuum*, 99: 76–79.

Zachariah SM, Lakshmy KS, and Thomas S (2024) Characterization techniques and fabrication methods used for nanocomposites. In: *Nanostructured Materials for Food Packaging Applications*, Jissy Jacob, Ilaria Cacciotti and Sabu Thomas (Eds.), Elsevier, pp. 69–81.

Zhang C (John), Ma Y, Zhang X, Abdolhosseinzadeh S, Sheng H, Lan W, Pakdel A, Heier J, and Nüesch F (2020) Two-dimensional transition metal carbides and nitrides (MXenes): synthesis, properties, and electrochemical energy storage applications. *EEM*, 3(1): 29–55.

Zhang X, Zheng J, and Ma Z (2019) Graphene-polymer composites for enhancing the mechanical properties. *IOP Conf Ser Mater Sci Eng*, 538(1): 012001

Zhao C, Li R, Fang B, Wang R, Liang H, Wang L, Wu R, Wei Y, Wang Z, Su Z, and Mo R (2023) 3D-printed stretchable conductive polymer composites with nano-carbon fillers for multifunctional applications. *CEST*, 1(2).

Zotti A, Zuppolini S, Borriello A, and Zarrelli M (2022) Polymer nanocomposites based on graphite nanoplatelets and amphiphilic graphene platelets. *Compos B Eng*, 246: 110223.

5 Optical and Photonic Properties in Polymer Nanocomposites

Deepika Sandil and Vishwa Pratap Singh

5.1 INTRODUCTION

In the last few decades, polymer nanocomposites have gained significant focus, preferably in the optical domain, because of their versatility in optical characteristics, leading to enhanced functional materials. The optical characteristics of the polymer itself, which include refractive index, transparency, and absorption spectra, form the foundation upon which the nanocomposite's properties are built. The embedding of nano-sized particles into the polymer matrix significantly enhanced these intrinsic properties, tailoring the material for specific optical applications.

Polymers, by nature, possess a wide range of optical characteristics, from high transparency and low-refractive index to various degrees of optical dispersion and birefringence. These properties make them ideal candidates for optical devices such as lenses, waveguides, and flexible displays. The polymer matrix's flexibility and processability also allow for the development of complex shapes and structures, further expanding their application potential in the optical domain. The introduction of nano-sized fillers in polymer matrices has potential in achieving the optimal output as these fillers interact with the polymer chains in ways that can either enhance or diminish the optical properties, depending on the intended application. For instance, polymers with naturally high transparency, such as polymethyl methacrylate (PMMA) or polyethylene terephthalate (PET), can be combined with nanoparticles to create composites with tailored optical transparency, refractive index, and light scattering properties (Jian et al., 2010; Kruthauser et al., 2006). This is achieved by optimizing the match between the refractive indices of the polymer and the nanoparticles, minimizing Rayleigh scattering, and ensuring uniform dispersion of the nanoparticles (Julien et al., 2019). Scheme 5.1 depicts the optical and photonic properties and applications of polymer nanocomposites.

Additionally, the refractive index of polymer nanocomposites can also be precisely engineered. Polymers with inherently low refractive indices can be modified by incorporating high-refractive index nanoparticles, such as titanium dioxide or zirconium oxide, to create composites with enhanced light manipulation capabilities (Cho et al., 2022; Tawfilas et al., 2024). This property is particularly valuable in photonic devices, where controlling light propagation and reflection is crucial. Conversely, for applications requiring low-refractive index materials, such as

DOI: 10.1201/9781003527602-5

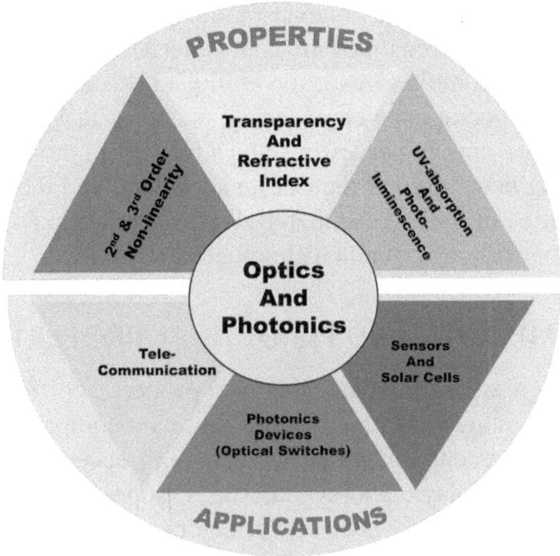

SCHEME 5.1 Optical and photonics properties and applications of polymer nanocomposites.

antireflective coatings or optical fibres, low-index fillers can be utilized to achieve the desired optical effect.

In 2010, Ma et al. reported some prime parameters that play a significant role in the overall performance of polymer composites. These parameters include the degree of dispersion of nano-sized particles, the interfacial coalition between matrix and nanoparticles and also the orientation of the nanoparticles (Ma et al., 2010). Considerably, the polymer's molecular structure, synthesis of polymer nanocomposite, and processing conditions are suggested to be an effective approach in controlling the dispersion of nanoparticles and enhancing the bonding at the interface, thereby affecting the optical homogeneity and reducing optical losses. Apart from this approach, surface functionalization of the nano-sized particles also effectively enhances their compatibility with the polymer, leading to better dispersion and more uniform optical properties, which is essential for high-performance optical applications (Fernandes et al., 2015).

The interaction between the polymer and nano-sized particles also leads to the emergence of unique optical phenomena, such as quantum confinement effects and plasmonic resonances. In cases where the nano-sized particles are of semiconducting or metallic nature, the polymer matrix can provide a stabilizing environment that allows these nano-sized particles to exhibit quantum confinement, resulting in discrete energy levels and altered absorption and emission properties. This effect is particularly useful in designing light-emitting diodes (LEDs) and photovoltaic cells. Plasmonic effects, which occur when metallic nanoparticles such as gold or silver are embedded in the polymer, lead to strong localized surface plasmon resonances, significantly enhancing light absorption and scattering at specific wavelengths. These effects can be harnessed in applications ranging from biosensing to photothermal therapies (Di Cianni et al., 2023).

In material science, the focus on polymer-based optical properties in nanocomposites continues to expand, driving innovations in material design and applications. The versatility of polymers, combined with the ability to precisely control their optical properties through nanocomposite engineering, opens up new possibilities in fields, namely optoelectronics, energy harvesting, and advanced photonics. Researchers are continually exploring new polymer matrices, nanofillers, and fabrication techniques to push the boundaries of what these materials can achieve, paving the way for the next generation of optical devices and technologies.

5.2 NANOCOMPOSITE AND ITS OPTICAL PROPERTIES

Composite materials have physical and chemical properties different from their primary components. Among different composites, nanocomposites have witnessed exceptional growth in recent times due to their high acceptance in diverse fields such as light-emitting displays, photovoltaic cells, supercapacitors, and opto-electronic devices (Shameem et al., 2021). The factors specifically dimension, shape, nature, effective surface area, interfacial bonding, and weight % of incorporated nanoparticles play a significant role in elevating optical properties of nanocomposites. Materials especially semiconducting and metals illustrate size-dependent optical properties at nano-dimension due to quantum confinement effects and the surface plasmon resonance phenomenon, respectively. Herein, we can take an example of a gold nanoparticle that changes its colour from green to orange as the size of the nanoparticle increased from 25 nm to 100 nm. Metal oxide-based nanocomposites especially TiO_2, ZrO_2, ZnO, and NiO have been researched well due to their extraordinary physical and optical properties (Cho et al., 2022; Liou et al., 2010; Salim and Tarabiah, 2023). In recent times, an investigation on nano-sized material such as reduced graphene oxide (rGO) has also been explored in detail, arguably due to its excellent optical and electronic behaviour (Sun et al., 2021).

In polymer nanocomposites, the weight % of incorporated nano-sized materials also plays a significant role in tailoring optical features of the composite. Musikhin et al. (2002) have carried out investigations on the luminescent properties of polymer nanocomposites using different nanomaterials such as Al_2O_3, Y_2O_3, SnSbO, and ZnO. In recent times, researchers have studied poly (vinyl alcohol) (PVA)-Gd_2O_3 polymer nanocomposites with variation in the concentration of Gd_2O_3 using different spectroscopy techniques. The UV–visible studies showed that with the increase in Gadolinium nanoparticle concentration, the optical bandgap decreases in the range of 5.78–4.86 eV (Madhuri et al., 2019). Apart from these optical behaviours, nano-sized particles in polymer nanocomposites can also produce nonlinear optical behaviour with the tuning of refractive index from positive to negative value.

5.3 MODULATION OF INTRINSIC OPTICAL PROPERTY: REFRACTIVE INDEX

The optical properties of materials, particularly refractive index modulation, are of paramount importance in various applications, ranging from optoelectronics to advanced photonic devices. The refractive index (n) is a fundamental optical property that explains the behaviour of light propagation through a material. Refractive indices of most common polymers fall in the range of 1.30–1.69 (Brandrup et al., 1999). However, these values of refractive index limit its applications in the field of image sensors, display devices, and others desiring the tailoring of polymers to produce high-refractive index ($n = 1.70$–2.50)-based functional polymers. The introduction of nano-sized particles into the conventional polymer matrices is an effective approach in preparing high refractive index functional polymers and can be termed as a polymer nanocomposite. Controlling and accurately measuring the refractive index in polymer nanocomposites is crucial for optimizing their performance in optical and photonic devices. Different applications have different requirements, for instance, in the case of optical waveguide, it is required to have a low refractive index, particularly in the range of 1.25–1.30, whereas in the case of filters or display surfaces, it is required to have high refractive index-based materials. In this section, we will discuss the engineering of the refractive index for the desired output.

Throughout the decades, high refractive index-based materials have been explored from fundamental as well as applied science due to their potential optical advancement. Inorganic materials such as metals, metal and nonmetal-based oxides, sulphides, or phosphides often display good mechanical properties, high thermal stabilities, and high refractive index, but they do suffer disadvantages in terms of high densities and low flexibility which limits its applications in the field of optical devices (Caseri, 2000). In contrast to these materials, organic polymers exhibit less weight, better flexibility, ease of fabrication, and good resistance, making it a better choice for optical applications but with a limitation in terms of value of refractive index (1.35–1.60) (Li et al., 2010; Wei et al., 2016). Nevertheless, researchers have shown that polymers with enhanced refractive index (n more than 1.7) can be produced with the incorporation of nano-sized fillers having high molar refraction and aromatic contents and low molar volume in accordance with Lorentz–Lorenz equation (Lorenz, 1880).

5.3.1 Nanocomposite Structures for Tunable Refractive Index

Refractive index of a polymer nanocomposite is definitely influenced by the choice of nano-sized particle and polymer matrix in terms of its refractive indices as well as volume fractions of respective components of the nanocomposite. Authors have reported a close approximation in the estimation of refractive index (n) of composite expressed by the equation:

$$n_{\text{composite}} = n_{\text{nanoparticle}} \times \phi_{\text{nanoparticle}} + n_{\text{polymer}} \times \phi_{\text{polymer}} \tag{5.1}$$

Here, $n_{\text{composite}}$, $n_{\text{nanoparticle}}$, and n_{polymer} represent the refractive indices of the resultant composite, nano-sized particle, and polymer matrix, respectively. The remaining two parameters $\phi_{\text{nanoparticle}}$ and ϕ_{polymer} represent as volume fractions of nanoparticles and polymer, respectively in above equation.

The selection of nanoparticles in designing a nanocomposite of definite-value refractive index is highly dominated by the concentration, dimension, and surface functionality of the nano-sized particles. The nanocomposite often suffers agglomeration of nanoparticles, preferably during synthesis time; thus, modification of these particle's surface is preferred. Numerous studies using oxide-based and sulphide-based nanomaterials, namely TiO_2, ZrO_2, ZnO, ZnS, FeS, and PbS, have been reported for the exploration of high refractive index polymer nanocomposites.

Nakayama et al. reported the linear increase in the refractive index of TiO_2-poly (bisphenol-A and epichlorohydrin) nanocomposite with the increase in the concentration of TiO_2 nanoparticles from 0 to 80 wt.% (Nakayama and Hayashi, 2007). In this study, the authors demonstrated that these composite materials can attain suitability as industrial high refractive index films and for optical devices. In another study, Tawfilas et al. reported the fabrication of transparent, high refractive-index polymer nanocomposite (Titanium dioxide-thermoplastic Polyurethane) with improved mechanical properties. Herein, authors have successfully increased the refractive index of the nanocomposite by 10.4% with 50 wt.% loading of titanium dioxide while maintaining its transparency (Tawfilas et al., 2024). Apart from well-studied TiO_2, zirconium oxide nanomaterial is also a preferable material towards the development of polymer nanocomposites in the field of optics. Cho et al. have done an interesting study on modified ZrO_2 towards the fabrication of highly transparent ZrO_2@SiO_2-PDMS nanocomposite film with tunable refractive indices (Cho et al., 2022).

Similarly, high refractive-indexed polymer nanocomposites, namely ZnS and PbS have also been discussed in the past for the potential applications in optical devices. Lü et al. reported the immobilization of thiol-capped ZnS nano-sized particles (2–5 nm in diameter) into the poly (urethane-methacrylate macromer) PUMM to produce transparent nanocomposite films (Lü et al., 2003). In this work, the authors studied the structural and optical properties of thiol-capped ZnS nanoparticles and its polymer nanocomposite using TEM, FTIR, TGA, and ellipsometer. The nanocomposite films showed good thermal stability along with a smooth surface, suggesting a uniform dispersion of the ZnS into the PUMM matrix. Further, the films also showed good optical transparency in the visible spectrum and tunable high refractive indices dependent on the concentration of ZnS nano-sized particles (Lü et al., 2003). In continuation, researchers have also reported a novel approach for the synthesis of high refractive index PbS/PTU polymer nanocomposite thin films for the application in optical coatings (Lü et al., 2005). In their work, researchers had exposed the Pb-containing precursors/polythiourethane (PTU) composite to H_2S gas to produce a transparent film.

5.3.2 Polymer Matrix and Refractive Index

Polymer modifications such as copolymerization or the addition of side chains are also one of the factors that have implications on the refractive index. These modifications change the polymer's density, molecular structure, and polarizability, thereby affecting the light interaction with the material. For instance, Tanaka et al. investigated the relationship between the refractive index and chemical structures of polyoctahedral oligomeric silsesquioxanes as an additive in PMMA matrix. In this investigation, the authors showed that less entanglement and small-scale interactions between the additives and polymer chains lower the packing and thus lead to the lowering of refractive index (Tanaka et al., 2010).

5.4 NONLINEAR OPTICAL PROPERTIES

In optics, linear and nonlinear characteristics of a material are intensity-independent and dependent prodigy respectively. However, nonlinear optics have gained significant attention due to its wide applications in the field of communication networks, optical signal processing, optical switching, storage and optical fibre (Singh and Singh, 2007). The interaction of a light wave with matter is generally expressed through induced polarization P. In response to less intense electromagnetic light waves, the polarization P is directly correlated with the strength of the applied field and the propagating electromagnetic wave exhibits time-varied phase and depends directly on the refractive index of the matter. However, the interaction of a highly intense electromagnetic wave with an isotropic nonconducting matter or medium often results in nonlinearity behaviour. In such interaction, the newly generated electromagnetic waves have variation not only in their frequency but also in their phase as well as other physical features. Fundamentally, the inception of nonlinearity in materials depends on the anharmonic motion of bound electrons under the influence of an applied electromagnetic field (Del Coso and Solis, 2004).

In this context, researchers have shown that the resultant total polarization P as induced by electric dipoles does not remain linear; however, it holds a more general relation as

$$P = \varepsilon_0\chi^{(1)}E + \varepsilon_0\chi^{(2)}\mathrm{E}^2 + \varepsilon_0\chi^{(3)}\mathrm{E}^3 + \ldots \tag{5.2}$$

Here, the parameter ε_0 signifies the permittivity of free space, and χ of different degrees signifies the respective order of susceptibility. In the above equation, the first term corresponds to the linear term and the second term onwards nonlinear term. In practice, $\chi^{(2)}$ and $\chi^{(3)}$ are very small in value and thus can be neglected. However, for a high intensity field, i.e. large value of electric fields, the susceptibility of higher orders becomes significant, and consequently, they lead to the occurrence of the new frequency components (Choudhary and Boyd, 2016).

Over the last few decades, a vast range of optical materials, namely polymers and their composites, nanostructured materials, bulk and doped glasses, have been well studied in the quest to produce new functional materials with enhanced properties. Amongst all materials, polymer nanocomposites have shown remarkable output in the field of waveguide sensors, e.m wave interference shielding and opto-electronic components (Soliman et al., 2020). The enhancement in the optical properties, specifically nonlinear optical properties, includes the modification in the nonlinear refractive index and third-order susceptibility. Studies have shown that these parameters are tunable and get tuned with the variation in the size or the nature of nano-sized fillers incorporated into the polymer matrix (Beecroft and Ober, 1997; Zihlif et al., 2012).

5.4.1 Nonlinear Behaviour in Polymer Nanocomposites

Nonlinear optical materials are highly applicable in the domains of optical switching and wavelength modulation due to their enhanced optical properties, typically third-order nonlinear characteristics. Nonlinear susceptibility (χ^3), nonlinear refractive index, and nonlinear absorption coefficient summarized the overall nonlinear characteristics of optical materials. Materials exhibiting particle's size in the dimension of nano-scale have shown the enhancement in optical characteristics attributed mainly to the quantum confinement effects.

5.4.1.1 Polymer Nanocomposites Based on Semiconductors

In polymer nanocomposites, nano-sized particles play a significant role in enhancing the nonlinear optical features of the composites. Semiconductor-based nanoparticles have always been experimented with in enhancing the optical features, as these features are directly dependent on the size and distribution of nano-sized particles. Technically, in semiconductors, one can observe strong quantum confinement as the size of particles approaches the Bohr radius with the appearance of the shift in the absorption spectrum, signifying a bandgap variation. Scheme 5.2 displays the quantum confinement effect observed in semiconductors-based materials.

Akimov et al. were the first to report work on semiconductor-based polymer nanocomposites to optimize optical media. In their work, researchers prepared cadmium sulphide (CdS) nano-sized particles in the range 2–50 nm. The author used different polymer matrices, namely poly (vinyl alcohol), poly (vinyl pyridine), and photographic gelatin and formed the composites with high CdS concentrations (up to half wt. %) without agglomeration. Lead sulphide (PbS)-based nanocomposites have also been explored by researchers to expose their nonlinearity properties. Asunskis et al. studied PbS-based polymer nanocomposites using different polymer matrices, namely PS, PE, poly (1-butene) and poly (1-decene) (Asunskis et al., 2008). Researchers reported the formation of transparent films of nanocomposites exhibiting sharp nonlinear absorption behaviour, which can be attributed to the existence of multiphoton absorption, trapped state, and free carrier absorption phenomenon.

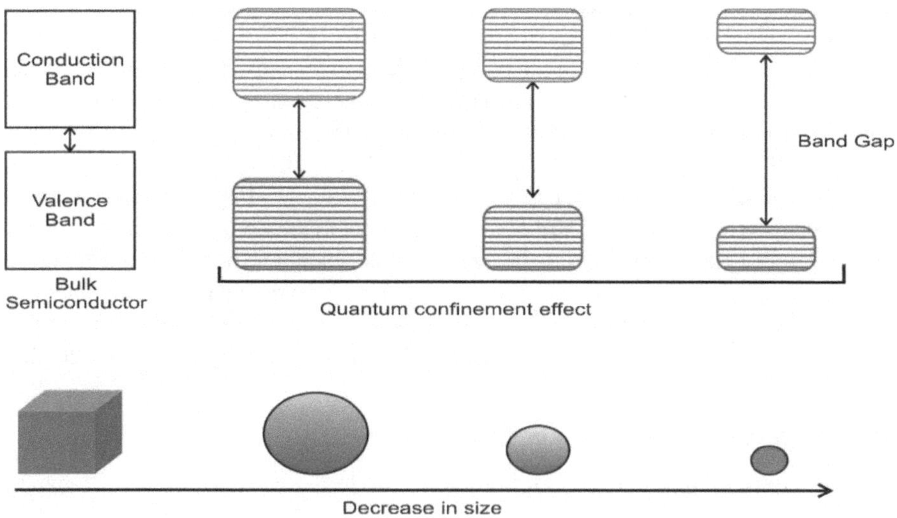

SCHEME 5.2 Quantum confinement effect in semiconductors.

Kulyk et al. did a study of nonlinear optical responses using spin-coated ZnO/PMMA thin films. In this work, they studied the second and third harmonic generation responses employing Z-scan technique at λ 1.064 micrometre and concluded that these higher-order nonlinear susceptibilities of as-prepared polymer nanocomposites were higher than bulk ZnO (Kulyk et al., 2009). Similarly, in 2010, Sreeja et al. employed ZnO-PMMA polymer nanocomposite thin films to study nonlinear optical absorption responses based on different nano-sized ZnO. Here, authors observed optical limiting type nonlinearity that occurred in response to 2-photon absorption in ZnO and reported an increased limit of efficiency with a decrease in energy bandgap. This response marked its application best in optical limiting. A numerous of works have been reported employing different ZnO-based polymer nanocomposites, which showed significant enhancement in their nonlinearity characteristics (Jeeju et al., 2014; Viswanath et al., 2014).

5.5 PHOTONICS

Photonics is one of the emerging disciplines of science and engineering where photons as the information carriers are studied to explore numerous technological innovations and incredible advancements in medical physics, quantum-based optics and metrology, imaging sciences, and optical communications. The ability to exert dependable control over light and its deterministically propagating waves in photonic materials is fundamental to these developments (Martin et al., 2001).

5.5.1 Fundamentals of Photonic Materials: Photonic Crystals and Bandgaps

Photonics is a study of science in terms of light generation, transmission, manipulation, and detection. Photonic materials are engineered to control the interaction of photons with matter, influencing properties such as speed, direction, and intensity of light. The fundamental characteristics of photonic materials include their refractive index, absorption coefficient, and scattering properties, all of which determine how light interacts with the material (Bhowmik et al., 2016). Photonic crystals, which have a periodic structure that influences photon mobility in a similar way to how electrons are affected by the periodic potential in a semiconductor crystal, are one of the most significant types of photonic materials (Huang et al., 2007). Photonic bandgaps are wavelength ranges in which light propagation is prohibited and can be produced using photonic crystals. This makes it possible to manipulate light output and propagation like never before.

To regulate light transmission, photonic crystals are devised with regularly shifting dielectric constants on the optical wavelength scale. The ability of photonic crystals to control the spontaneous emission processes of emitters placed within them is their most intriguing aspect (Comoretto, 2015). For photonic crystals, the dispersion relation displays the emergence of a photonic bandgap, in which frequencies falling inside the gap have no accessible states. As a result, extraordinary control over spontaneous emission is possible near the bandgap, where the density of photon states can be zero—a feature that was previously considered as unchangeable. The formation of photonic bandgaps can be understood by examining the Bragg's condition for constructive interference. When the periodicity of the photonic crystal matches half the wavelength of light, Bragg reflection occurs, leading to the formation of a bandgap (Zhao et al., 2020). Mathematically, this condition is expressed as

$$m\lambda = 2d\sin\theta - \tag{5.3}$$

where m signifies the order of the reflection, λ as the wavelength of light, d as the spacing between the planes in the crystal, and θ as the angle of incidence. Materials such as silicon (Si) and gallium arsenide (GaAs) are commonly used in fabricating photonic crystals due to their high refractive indices, which enhance bandgap formation (Bhowmik et al., 2016). One-dimensional photonic crystals are traditionally devised using techniques like molecular beam epitaxy and chemical vapour deposition, resulting in alternating layers with distinct refractive indices. Two-dimensional crystals, fabricated through dry reactive ion etching or wet electrochemical etching, offer precise control over nanostructures. For three-dimensional photonic crystals, advanced methods like colloid self-assembly and laser microfabrication are employed, despite challenges in achieving full electromagnetic wave control. These fabrication techniques and material choices are critical for designing photonic crystals with specific optical properties and applications.

5.5.2 Photonic Metamaterials and Photonic Bandgap Materials

Over the decades, the modern optics and photonics field have completely transformed with the advancement of metamaterial technology and has made it possible to design material characteristics in almost infinite ways and create completely new linear and nonlinear optical properties along with novel capabilities (Litchinitser and Shalaev, 2008). Photonic metamaterials are a broad class of micro- or nanostructures made up of specially designed building blocks that are ideally considerably smaller than the wavelength of light, allowing for dense packing into a material that works (Soukoulis and Wegener, 2011). Metamaterials, initially spurred by "a quest for a superlens" (Pendry, 2000) have given rise to an array of novel phenomena and possible uses, including spectrally selective filters, subwavelength waveguides, negative refraction, total external reflection, nanocircuits, antennas, cloaking devices, wave concentrators, and rotators (Smolyaninov et al., 2008; Engheta, 2007; Litchinitser and Shalaev, 2008).

Typically, two parameters are used to describe how electromagnetic waves interact with materials: the dielectric permittivity (ε) and the magnetic permeability (μ), both having real and imaginary components. Interestingly, the refractive index (n) of a metamaterial is expressed as

$$n = \pm\sqrt{\varepsilon\mu} \tag{5.4}$$

Recent advancements in nanofabrication have demonstrated that, with appropriate material engineering, any combination of ε and μ can theoretically be produced.

Such engineering is made possible in metamaterials by "meta-atoms," which stand in for normal atoms or molecules. Engineered structures called meta-atoms are tiny in relation to the wavelength of incoming radiation but much larger than natural atoms. Meta-atoms' shape, size, and other attributes can be altered to precisely construct and adjust their electric and magnetic properties. When it comes to electromagnetic wave interactions, meta-atoms may differ greatly from normal atoms (Landau et al., 1984; Litchinitser and Shalaev, 2008).

As stated earlier, photonic metamaterials have opened new avenues for manipulating light in ways that were previously thought impossible. These materials also exhibit properties such as negative refraction, which occurs when the refractive index is negative, causing light to bend in the opposite direction compared to normal materials. This phenomenon can be described by Snell's law for negative refraction:

$$n_1 \sin\theta_1 = -n_2 \sin\theta_2 \tag{5.5}$$

n_1 and n_2 signify as refractive indices of the two media, and θ_1 and θ_2 as the angles of incidence and refraction, respectively. The ability to achieve negative refraction enables the development of superlenses, which can overcome the diffraction limit of conventional lenses, allowing for imaging with resolution beyond the wavelength of light (Zhang and Liu, 2008).

The configuration of photonic crystals is inspired by the electronic analogue of controlling the flow of electrons in semiconductor crystals. By constructing photonic crystals with periodically varying dielectric constants, it is possible to control the transport of light through the material. One of the most exciting aspects of photonic crystals is their deterministic control over the spontaneous emission properties of an emitter embedded within them. This control is achieved through the modification of the photonic density of states (DOSs), which can be designed to have zero values at specific bandgaps, thereby providing exceptional control over the emission properties.

The DOSs in a photonic crystal can be significantly altered compared to conventional materials. The DOS, which represents the number of photonic states available at each energy level, is zero within the photonic bandgap. This results in the suppression of spontaneous emission for emitters with transition energies within the bandgap. The modification of the DOS can be represented mathematically as

$$\rho(\omega) = \frac{\omega^2}{\pi^2 c^3} \sum_n \delta(\omega - \omega_n) - \tag{5.6}$$

where $\rho(\omega)$ is the photonic DOSs, ω is the angular frequency, and ω_n are the resonant frequencies of the photonic modes.

5.6 POLYMER NANOCOMPOSITES AS PHOTONIC MATERIALS

Polymer-based materials are promising materials for photonics–photon-based research and technology. In recent times, the polymeric materials have also discovered their uses for biophotonics, a discipline that uses light for optical diagnostics and light-guided and activated therapy, as well as for nano-photonics, which uses the fusion of photonics and nanotechnology. By embedding nanoparticles such as gold, silver, or carbon nanotubes into polymer matrices, researchers have developed materials with tailored optical characteristics, including enhanced photoluminescence (PL), nonlinear optical behaviour, and tunable refractive indices. For instance, an inorganic quantum dots dispersed into a polymer-bed have shown improved carrier mobility and tunable photo-response, and increased photogeneration efficiency (Prasad, 2004). These enhanced properties have proven valuable for numerous opto-electronic applications.

Nanotube–polymer composites, for example, have shown promise for ultrafast photonics due to the high third-order nonlinear optical susceptibility of carbon nanotubes (Hasan et al., 2009). These composites can be used in applications such as optical switches, modulators, and sensors. The integration of nanoparticles into polymers allows for the design of materials with specific optical responses, enabling the development of advanced photonic devices with improved performance and new functionalities.

The optical properties of polymer nanocomposites as discussed earlier get influenced by factors such as the size, shape, and distribution of the nanoparticles, as well as the interaction between the nanoparticles and the polymer matrix. Effective medium theories, such as the Maxwell–Garnett theory, are often used to predict the optical behaviour of these composites. The real and imaginary parts of the effective refractive index can be calculated by considering the contributions of both the polymer matrix and the embedded nanoparticles.

5.6.1 Role of Polymer Nanocomposites in Photonics

Polymer nanocomposites play a crucial role in the field of photonics due to their ability to combine the advantageous properties of polymers with the unique optical characteristics of nanoparticles. This synergy results in materials that are not only mechanically robust but also exhibit enhanced optical functionalities. Here, we delve into the role and significance of polymer nanocomposites in photonics, supported by detailed insights from research articles.

5.6.2 Enhanced Optical Properties

One of the primary advantages of polymer nanocomposites is their ability to enhance optical properties through the inclusion of nanoparticles. For example, metal nanoparticles such as gold and silver can significantly improve the optical absorption and scattering properties of the polymer matrix. These enhancements are attributed to the localized surface plasmon resonances of the metal nanoparticles, which can be tuned by adjusting their size, shape, and distribution within the polymer matrix.

According to Bockstaller and Thomas (2003), the incorporation of gold nanocrystals within a block copolymer matrix led to significant alterations in the dielectric function of the composite, enhancing its reflective properties. The study demonstrated that by preferentially sequestering gold nanocrystals within the polystyrene (PS) domains of a poly(styrene-b-ethylene/propylene) copolymer, the nanocomposite exhibited high rejection levels for incident light, a critical requirement for photonic applications (Bockstaller and Thomas, 2003).

5.6.3 Nonlinear Optical Behaviour

Polymer nanocomposites also exhibit remarkable nonlinear optical properties, making them suitable for applications in ultrafast photonics. Nanotube–polymer composites, for instance, have shown promise in this area due to the high third-order nonlinear optical susceptibility of carbon nanotubes. These composites can be employed in optical switches, modulators, and sensors, which are essential components in modern photonic devices (Hasan et al., 2009).

Hasan et al. (2009) discussed the potential of nanotube–polymer composites for ultrafast photonics, highlighting their suitability for applications requiring rapid response times and high efficiency. The integration of carbon nanotubes into polymer matrices allows for the design of materials with specific optical responses, enabling the development of advanced photonic devices with enhanced performance.

5.6.4 Tunable Refractive Indices

Another significant aspect of polymer nanocomposites is their tunable refractive indices. By varying the type and concentration of nanoparticles embedded within the polymer matrix, researchers can tailor the refractive index of the composite material to suit specific applications. This tunability is particularly useful in the design of photonic crystals and waveguides, where precise control over the refractive index is necessary to achieve the desired optical effects.

The effective medium theory (EMT) is often used to model and predict the optical behaviour of polymer nanocomposites. The Maxwell–Garnett theory, for instance, provides a means to calculate the effective dielectric constant of the composite based on the properties of the individual components. This approach was validated by Bockstaller and Thomas (2003), who found that the Maxwell–Garnett theory accurately described the optical properties of their gold nanocrystal-filled block copolymer composites (Bockstaller and Thomas, 2003).

5.7 PL AND FLUORESCENCE PROPERTIES OF POLYMER NANOCOMPOSITES

In photonics, understanding the mechanism of PL is crucial for designing and optimizing various optical devices. Photoluminescent materials are used in LEDs, laser diodes, dye-sensitized solar cells, and other opto-electronic components. The efficiency and stability of these devices depend significantly on the PL properties of the materials used. PL spectroscopy is a powerful tool for characterizing these materials. It provides insights into the electronic band structure, defect states, and carrier dynamics, which are essential for tailoring the optical properties of photonic devices. By studying the PL spectrum, researchers can identify the presence of impurities, defects, and other factors that influence the performance of photonic devices.

Fluorescence behaviour in polymer matrices is a significant area of study in material science and photonics due to its applications in sensing, imaging, and light-emitting devices. Polymer matrices provide a versatile and tunable environment for fluorescent molecules, offering unique advantages such as mechanical flexibility, processability, and the ability to form thin films and fibres. Fluorescence in polymer matrices involves the absorption of light by fluorescent molecules (fluorophores) embedded within the polymer. Upon absorption of a photon, an electron in the fluorophore is excited from the ground state (S_0) to an excited state (S_1). This excitation is followed by a nonradiative relaxation to the lowest vibrational level of the excited state. The electron then returns to the ground state by emitting a photon, a process known as fluorescence. This emitted photon has a longer wavelength than the absorbed photon, in accordance with a phenomenon known as the Stokes shift.

5.7.1 Photoluminescence Mechanism

PL is a crucial phenomenon, where a material absorbs photons and subsequently re-emits them as light. This process is integral in understanding and exploring new size and surface-based electronic and optical features exhibited by materials, particularly semiconductors. The mechanism of photoluminescence involves several stages, including excitation, relaxation, and emission, which can be described in terms of interactions between electrons and their corresponding energy states.

5.7.2 Absorption and Excitation

The PL process begins with the absorption of photons by the material, which elevates electrons from the low-energy ground state (S_0) to high-energy excited state (S_1). This transition dominates only when the incident photon E has energy greater than or matches the bandgap between the ground and excited states. This relationship is described by

$$E_{\text{photon}} = hv = E_{S1} - E_{S0} \tag{5.7}$$

h signifies as Planck's constant, v as frequency of the incident light, E_{S1} as the energy of the excited state, and E_{S0} as the energy of the ground state.

In semiconductors, this excitation promotes charge carriers' electrons from the state of the valence band to the state in the conduction band, creating electron–hole pairs. The excited electrons occupy higher energy states, while the corresponding holes remain in the lower energy states, leading to the formation of excitons.

5.7.3 Photoluminescence and Fluorescence Behaviour in Polymer Matrices

The photoluminescence of polymer nanocomposites depends on the type of polymer and nanoparticles used in the composite. In 2004, Pucci et al. demonstrated the synthesis and formation of polymer nanocomposite based on gold (Au) nanoparticles incorporated into ultrahigh molecular weight polyethylene matrix. The authors had carried out the different optical studies and reported the existence of strong enhanced PL ($\lambda = 288$ nm) (Pucci et al., 2004). One can also enhance the luminescent properties of the polymers by incorporating the inorganic nano-sized particles into conjugated polymers (Guo et al., 2007). A lot of research in developing such functional materials for the development of highly efficient photovoltaic devices has been reported (Xu et al., 2005; Yu et al., 2005). For instance, Zou et al. have reported the studies in exploring the new approach to enhance the PL efficiency of a strong luminescent material polyfluorene. The authors have employed n-type ZnO nanoparticles into the conjugated polyfluorene matrix in a controlled manner to produce high-performance voltaic devices. The studies revealed the strong enhancement in luminescent characteristics with the incorporation of 10% ZnO into the polymer matrix (Zou et al., 2011).

The polymer matrix significantly influences the fluorescence behaviour of the embedded fluorophores. Factors such as the polarity, rigidity, and molecular interactions within the polymer can alter the fluorescence intensity and lifetime. For instance, in rigid matrices, the fluorophores experience less rotational and vibrational relaxation, leading to higher fluorescence quantum yields. Conversely, in more flexible or fluid environments, nonradiative decay processes may be enhanced, reducing fluorescence efficiency. Polymer matrices can also affect the photostability of fluorophores. Some polymers can protect the fluorophores from quenching agents, such as oxygen, which can otherwise reduce fluorescence by promoting nonradiative relaxation pathways (Wöll, 2014). Kuila et al. investigated the luminescent properties of poly(3-hexyl thiophene)/hexadecyl amine-capped Ag nanocomposites in solution and as a thin film. Their studies revealed that the intensity of the polymer nanocomposite decreases in solution with the increase in the concentration of Ag nanoparticles, whereas on investigating the nanocomposite as a thin film, enhanced fluorescence intensity was observed (Kuila et al., 2007). In another study, the authors have explored a highly sensitive fluorescence sensor using polyvinylpyrrolidone (PVP)-GO nanocomposites towards the detection of toxic compounds, specifically nitrite ions. Their study revealed that PVP-GO nanocomposite depicts good sensitivity along with good selectivity and reproducibility and concluded its significant contribution towards the development of fluorescence sensors (Hua, 2017). Nevertheless, in the search of new functional materials in the field of optics, carbon-based nanocomposites have also been investigated owing to their interesting PL features (De, 2019). In the past decade, many researchers have studied the different parameters contributing to PL characteristics of carbon dots, namely the size of the dots and the functional groups attached to the surface (De et al., 2017; Zhu et al., 2015). Zhou et al. produced light-emitting, amphibious nanocomposite films by blending carbon dots with PMMA in dichloromethane and PVA in water, followed by solvent evaporation. The addition of carbon dots imparts strong photoluminescent (PL) properties to the polymer films while maintaining their transparency (Zhou et al., 2013).

5.8 PHOTONICS APPLICATIONS

5.8.1 Optical Waveguides and Fibres

Optical waveguides and fibres are critical components in photonics, enabling efficient light transmission over short and long distances. These structures guide light by confining it within a core material, surrounded by a cladding with a lower refractive index, thus enabling minimal signal loss due to total internal reflection.

Polymer nanocomposite waveguides offer several advantages, including ease of fabrication through techniques like soft lithography and roll-to-roll processing. They are also compatible with flexible substrates, making them ideal for integration into wearable photonic devices and flexible electronics. These materials are particularly advantageous in the development of compact and lightweight photonic devices, essential for telecommunications, medical diagnostics, and environmental monitoring. The ability to tailor the optical and mechanical properties through the choice of polymer and nanoparticle combination enables the creation of customized solutions for specific applications

5.8.2 Polymer Nanocomposites in Optical Waveguides

Polymer nanocomposites have emerged as promising materials for optical waveguides and fibres due to their unique combination of flexibility, processability, and tunable optical properties. These composites consist of a polymer matrix embedded with nanoparticles, such as silica or metal oxides, which enhance the optical performance and mechanical strength of the waveguide.

Several polymer nanocomposites have been studied for their effectiveness in optical applications:

- PMMA with silica nanoparticles enhances the refractive index contrast ($\Delta n = 0.02$–0.04), improving light confinement and reducing propagation losses (0.2–0.5 dB/cm).
- Polyimide with titanium dioxide (TiO_2) nanoparticles provides high thermal stability (up to 400°C) and improved optical properties, suitable for high-temperature applications.
- PS with zinc oxide (ZnO) nanoparticles offers enhanced UV resistance (UV absorption at 370 nm) and optical clarity (transmittance > 90%), ideal for applications requiring high transparency.
- PVA with silver (Ag) nanoparticles exhibits strong plasmonic effects, useful in sensing applications due to its enhanced sensitivity to changes in the environment (sensitivity factor ~1.5).
- Polycarbonate with gold (Au) nanoparticles provides excellent mechanical strength (Young's modulus increases by 20%) and enhanced nonlinear optical properties (nonlinear refractive index ~$10^{-7}\,cm^2/W$), suitable for advanced photonic applications.

5.9 CONCLUSION AND FUTURE SCOPE

This chapter presents a brief overview of different optical and photonic aspects of polymer nanocomposites. The linear and nonlinear optical properties of polymer nanocomposites include transparency, linear refractive index, absorption, and second- and third-order susceptibilities, respectively. These properties are remarkably influenced by the kind of polymer or nanoparticles employed, the size of nanoparticles (specifically when quantum dots are involved), scattering, agglomeration, and processing conditions. The basic mechanism of each property has been introduced.

The modulation of the refractive index of incorporated nanoparticles leads to the development of high-performance polymer nanocomposites, specifically in waveguides, antireflected coatings, lenses, and optical filters. With the incorporation of high refractive-index nanomaterials namely ZnO, TiO_2, ZrO_2, AL_2O_3, PbS, and ZnS into transparent polymers, a transparent functionalized nanocomposite can be achieved. Polymer nanocomposites developed using gold and silver nanoparticles exhibit high nonlinear coefficients and can be employed in the domain of frequency modulation and ultrafast photonics. Polymer nanocomposites also offer significant scope in the development of organic LED owing to its enhanced luminescence characteristics. In this context, quantum dots embedded in a polymer matrix can play a significant role in producing highly efficient and brighter light-emitting devices. Further, polymer and copolymer systems which exhibit long-range periodicity and dielectric constants can significantly contribute to the fabrication of photonic crystals.

Thus, owing to interesting and tunable optical characteristics, nano-dimensional structures-based polymer nanocomposites always offer an extensive scope of research due to their potential applications. This further holds significant future in exploring functionalized materials for advancements in producing flexible, economical, and sustainable optical devices.

REFERENCES

Asunskis, D. J., Bolotin, I. L., & Hanley, L. (2008). Nonlinear optical properties of PbS nanocrystals grown in polymer solutions. *J. Phys. Chem. C*, *112*:9555–9558.

Brandrup, J., Immergut, E. H., Grulke, E. A., Abe, A., & Bloch, D. R. (Eds.). (1999). *Polymer Handbook*. Wiley.

Kulyk, B.,Sahraoui, B., Krupka, O., Kapustianyk, V., Rudyk, V., Berdowska, E., Tkaczyk, S., & Kityk, I. (2009). Linear and nonlinear optical properties of ZnO/PMMA nanocomposite films. *J. Appl. Phys.* *106*(9):093102–093102-6.

Beecroft, L. L., & Ober, C. K. (1997). Nanocomposite materials for optical applications. *Chem. Mater.* *9*(6):502–517.

Bhowmik, K., Peng, G. D., Ambikairajah, E., & Rajan, G. (2016). Photonic materials for sensing, biosensing and display devices. *Springer Ser. Mater. Sci. 229*.

Bockstaller, M. R., & Thomas, E. L. (2003). Optical properties of polymer-based photonic nanocomposite materials. *J. Phys. Chem. B 107*(37):10017–10024.

Brandrup, J., Immergut, E. H., & Grulke, E. A. (1999). *Polymer Handbook*, 4th ed. John Wiley & Sons.

Caseri, W. (2000). Nanocomposites of polymers and metals or semiconductors: historical background and optical properties. *Macromol. Rapid Commun. 21*:705–722.

Cho, H., Lee, D., Hong, S., Kim, H., Jo, K., Kim, C., & Yoon, I. (2022). Surface modification of ZrO2 nanoparticles with TEOS to prepare transparent ZrO2@SiO2-PDMS nanocomposite films with adjustable refractive indices. *Nanomaterials 12*(14):2328.

Choudhary, S., & Boyd, R. W. (2016). Tutorial on nonlinear optics. In D. Faccio, J. Dudley and M. Clerici (Eds.), *Frontiers in Modern Optics* (pp. 31–76). IOS Press.

Comoretto, D. (2015). Organic and hybrid photonic crystals. In *Organic and Hybrid Photonic Crystals*. Springer.

De, B. (2019). Carbon dots and their polymeric nanocomposites. In Niranjan Karak (Ed.), *Nanomaterials and Polymer Nanocomposites: Raw Materials to Applications* (pp. 217–260). Elsevier.

De, B., Kuila, T., Kim, N. H., & Lee, J. H. (2017). Carbon dot stabilized copper sulphide nanoparticles decorated graphene oxide hydrogel for high performance asymmetric supercapacitor. *Carbon, 122:* 247–257.

Del Coso, R. L., & Solis, J. (2004). Relation between nonlinear refractive index and third-order susceptibility in absorbing media. *J. Opt. Soc. Am. B 21*(3): 640–644.

Di Cianni, W., de la Mata, M., Delgado, F. J., Hernández-Saz, J., Herrera, M., Molina, S. I., Giocondo, M., & Sanz de León, A. (2023). Polymer nanocomposites for plasmonics: *In situ* synthesis of gold nanoparticles after additive manufacturing. *Polym. Test. 117*: 107869.

Engheta N. (2007). Circuits with light at nanoscales: optical nanocircuits inspired by metamaterials. *Science 317*(5845): 1698–1702

Fernandes, R. M. F., Abreu, B., Claro, B., Buzaglo, M., Regev, O., Furó, I., & Marques, E. F. (2015). Dispersing carbon nanotubes with ionic surfactants under controlled conditions: comparisons and insight. *Langmuir 31*(40):10955–10965.

Guo, Z., Wei, S., Shedd, B., Scaffaro, R., Pereira, T., & Hahn, H. T. (2007). Particle surface engineering effect on the mechanical, optical and photoluminescent properties of ZnO/vinyl-ester resin nanocomposites. *J. Mater. Chem. 17*(8):806–85.

Hasan, T., Sun, Z., Wang, F., Bonaccorso, F., Tan, P. H., Rozhin, A. G., & Ferrari, A. C. (2009). Nanotube–polymer composites for ultrafast photonics. *Adv. Mater. 21*(38-39):3874–3899.

Hua, T. I. (2017). *Polyvinylpyrrolidone, graphene oxide and their composites as potential fluorescence sensing materials for nitrate and nitrite ions*. Doctoral dissertation, Universiti Teknologi Malaysia.

Huang, C. K., Chan, C. H., Chen, C. Y., Tsai, Y. L., Chen, C. C., Han, J. L., & Hsieh, K. H. (2007). Rapid fabrication of 2D and 3D photonic crystals and their inversed structures. *Nanotechnology 18*(26), 265305.

Jeeju, P. P., Jayalekshmi, S., & Chandrasekharan, K. (2014). Nonlinear optical properties of ZnO/poly (vinyl alcohol) nanocomposite films. In: *Optoelectronics Materials and Thin Film, AIP Conference Proceedings, vol.* 1576 (pp. 163–166). Kochi, Kerala, India.

Jian, H. Y., Alex, J. H., & Gregory, C. R. (2010). *Novel Transparent PMMA Composites for Optical Tagging. Army Research Laboratory Report ARL-TR-5354.*

Julien, L., Jose-Marie, L, -C., Laurent, B., Hélène, G., & Maud, S. (2019). Transparent polymer nanocomposites: an overview on their synthesis and advanced properties. *Prog. Polym. Sci. 89*:133–158.

Krauthauser, C., Deitzel, J. M., O'Brien, D. & Hrycushko, J. (2006). Optical properties of transparent resins with electrospun polymer nanofibers. In: D. H. Reneker, H. Fong (Eds.), *Polymeric Nanofibers* (pp. 353–369). ACS Publication.

Kuila, B. K., Garai, A., & Nandi, A. K. (2007). Synthesis, optical, and electrical characterization of organically soluble silver nanoparticles and their poly(3-hexylthiophene) nanocomposites: enhanced luminescence property in the nanocomposite thin films. *Chem. Mater. 19*(22):5443–5452.

Landau, L. D., Lifshitz, E. M., & Pitaevskii, L. P. (1984). Electrodynamics of Continuous Media. Buterworth.

Li, C., Li, Z., Liu, J. -G., Zhao, X. -J., Yang, H. -X., & Yang, S. -Y. (2010). Synthesis and characterization of organo-soluble thioether-bridged polyphenylquinoxalines with ultra-high refractive indices and low birefringences. *Polymer 51*: 3851–3858.

Liou, G. -S., Lin, P. -H., Yen, H. -J., Yu, Y. -Y., Tsai, T. -W., & Chen, W. -C. (2010). Highly flexible and optical transparent 6F-PI/TiO_2 optical hybrid films with tunable refractive index and excellent thermal stability. *J. Mater. Chem. 20*(3):531–536.

Lorenz L. (1880) Ueber die Refractionsconstante. *Ann. Phys. 247*:70–103.

Lü, C., Cui, Z., Wang, Y., Li, Z., Guan, C., Yang, B., & Shen, J. (2003). Preparation and characterization of ZnS-polymer nanocomposite films with high refractive index. *J. Mater. Chem. 5*(9):2189–2195.

Lü, C., Guan, C., Liu, Y., Cheng, Y., & Yang, B. (2005). PbS/polymer nanocomposite optical materials with high refractive index. *Chem. Mater. 17*(9):2448–2454.

Ma, P. -C., Siddiqui, N. A., Marom, G., & Kim, J. -K. (2010). Dispersion and functionalization of carbon nanotubes for polymer-based nanocomposites: a review. *Compos. A: Appl. Sci. Manuf. 41*(10): 545–567.

Madhuri, S. N., Hemalatha, K. S., & Rukmani, K. (2019). Preparation and investigation of suitability of gadolinium oxide nanoparticle doped polyvinyl alcohol films for optoelectronic applications. *J. Mater. Sci.: Mater. Electron. 30*(9):9051–9063.

Martin, J., Kiesow, A., Heilmann, A., & Wannemacher, R. (2001). Laser microstructuring and scanning microscopy of plasmapolymer–silver composite layers. *Appl. Opt. 40*(31), 5726–5730.

Musikhin, S., Bakueva, L., Sargent, E. H., & Shik, A. (2002). Luminescent properties and electronic structure of conjugated polymer-dielectric nanocrystal composites. *J. Appl. Phys. 91*(10 I): 6679–6683.

Nakayama, N., & Hayashi, T. (2007). Preparation and characterization of TiO2 and polymer nanocomposite films with high refractive index. *J. Appl. Polym. Sci. 105*(6): 3662–3672.

Litchinitser, N. M., & Shalaev, V. M. (2008). Photonic metamaterials. *Laser Phys. Lett. 5*(6), 411.

Prasad, P. N. (2004). Polymer science and technology for new generation photonics and biophotonics. *Curr. Opin. Solid State Mater. Sci. 8*(1), 11–19.

Pendry, J. B. (2000). Negative refraction makes a perfect lens. *Phys. Rev. Lett. 85*: 3966.

Pucci, A., Tirelli, N., Willneff, E. A., Schroeder, S. L. M., Galembeck, F., & Ruggeri, G. (2004). Evidence and use of metal-chromophore interactions: Luminescence dichroism of terthiophene-coated gold nanoparticles in polyethylene oriented films. *J. Mater. Chem. 14*(24):3495–3502.

Salim, E., & Tarabiah, A. E. (2023). The influence of NiO nanoparticles on structural, optical and dielectric properties of CMC/PVA/PEDOT:PSS nanocomposites. *J. Inorg. Organomet. Polym. Mater. 33*(6):1638–1645.

Shameem, M. M., Sasikanth, S. M., Annamalai, R., & Raman, R. G. (2021). A brief review on polymer nanocomposites and its applications. *Mater. Today Proc. 45*:2536–2539.

Singh, S. P., & Singh, N. (2007). Nonlinear effects in optical fibers: origin, management and applications. *Prog. Electromag. Res. 73*:249–275.

Smolyaninov, I. I., Hung, Y. J., & Davis, C. C. (2007). Electromagnetic cloaking in the visible frequency range. *Opt. Lett. 33*:1342–1344.

Soliman, T. S., Vshivkov, S. A., & Elkalashy, S. I. (2020). Structural, linear and nonlinear optical properties of Ni nanoparticles – polyvinyl alcohol nanocomposite films for optoelectronic applications. *Opt. Mater. 107:*110037.

Soukoulis, C. M., & Wegener, M. (2011). Past achievements and future challenges in the development of three-dimensional photonic metamaterials. Nat. Photon. *5*(9):523–530.

Sun, X., Huang, C., Wang, L., Liang, L., Cheng, Y., Fei, W., & Li, Y. (2021). Recent progress in graphene/polymer nanocomposites. *Adv. Mater. 33*(6):2001105.

Tanaka, K., Adachi, S., & Chujo, Y. (2010). Side-chain effect of octa-substituted POSS fillers on refraction in polymer composites. *J. Polym. Sci. A Polym. Chem. 48*(24):5712–5717.

Tawfilas, M., Bartolini Torres, G., Lorenzi, R., Saibene, M., Mauri, M., & Simonutti, R. (2024). Transparent and high-refractive-index titanium dioxide/thermoplastic polyurethane nanocomposites. ACS Omega *9*(27):29339–29349.

Viswanath, V., Beenakumari, C., & Muneera, C. I. (2014). ZnO-PVA nanocomposite films for low threshold optical limiting applications. In: *Light and Its Intraction with Matter, AIP Conference Proceedings*, vol. 1620 (pp. 604–610). Kerala, India.

Wei, Q., Pötzsch, R., Liu, X., Komber, H., Kiriy, A., Voit, B., Will, P. -A., Lenk, S., & Reineke, S. (2016). Hyperbranched polymers with high transparency and inherent high refractive index for application in organic light-emitting diodes. *Adv. Funct. Mater. 26*:2545–2553.

Wöll, D. (2014). Fluorescence correlation spectroscopy in polymer science. *RSC Adv.* *4*(5): 2447–2465.

Zhang, X., & Liu, Z. (2008). Superlenses to overcome the diffraction limit. *Nat. Mater.* *7*(6): 435–441.

Xu, J., Shi, W., Gong, M., Yu, F., & Yan, L. (2005). Preparation of poly(methyl methacrylate-*co* -maleic anhydride)/SiO_2-TiO_2 hybrid materials and their thermo- and photodegradation behaviors. *J. Appl. Polym. Sci.* *97*(4):1714–1724.

Yu, D., An, J. H., Bae, J. Y., Ahn, S. D., Kang, S., & Suh, K. S. (2005). Preparation of titanium dioxide/poly(methyl methacrylate- co - n -butyl acrylate- co -methacrylic acid) hybrid composite particles via emulsion polymerization. *J. Appl. Polym. Sci.* *97*(1):72–79.

Zhao, F., Guo, Y., Zhou, X., Shi, W., & Yu, G. (2020). Materials for solar-powered water evaporation. *Nat. Rev. Mater.* *5*(5):388–401.

Zhou, L., He, B., & Huang, J. (2013). Amphibious fluorescent carbon dots: one-step green synthesis and application for light-emitting polymer nanocomposites. *Chem. Commun.* *49*(73):8078–8080.

Zhu, S., Song, Y., Zhao, X., Shao, J., Zhang, J., & Yang, B. (2015). The photoluminescence mechanism in carbon dots (graphene quantum dots, carbon nanodots, and polymer dots): current state and future perspective. *Nano Res.* *8*(2):355–381.

Zihlif, A., Faduos, A. S., & Ragosta, G. (2012). Optoelectrical properties of polymer composite. *J. Thermoplast. Compos. Mater.* *26*(9):1180–1191.

Zou, J. P., Le Rendu, P., Musa, I., Yang, S. H., Dan, Y., That, C. T., & Nguyen, T. P. (2011). Investigation of the optical properties of polyfluorene/ZnO nanocomposites. *Thin Solid Films* *519*(12):3997–4003.

6 Recent Advances in Polymer Nanocomposites and Their Impact on the Electronic and Electrical Industry

Shashank Sharma and Kalpana Singh

6.1 INTRODUCTION

Nanopolymer composites are novel materials that have garnered considerable interest in recent years due to their distinctive characteristics and extensive array of uses. Advances in nanotechnology and materials science have accelerated the field's expansion. The change in nanoparticle dispersion and interaction within the polymer matrix has opened up new opportunities. The dispersion of nanoparticles inside a polymer lattice results in composites that display heightened mechanical strength, customized characteristics suitable for certain applications, and enhanced overall performance. As a result, scientists have created sophisticated methods to make it easier to produce nanopolymers. Gaining a comprehensive understanding of the scientific principles underlying nanopolymer composites is crucial in order to fully exploit their capabilities. Each stage in the preparation process, from the creation of the polymer lattice to the distribution of nanoparticles, is vital in establishing the ultimate characteristics of the composite material. Furthermore, the interaction between the polymer and nanoparticles is crucial for attaining the best possible performance. Alongside the increased adoption of electronic gadgets, the electronics and electrical sectors have seen tremendous advancements in recent years. This has resulted in the emergence of a new category of pollution known as electromagnetic interference (EMI). Electromagnetic coupling, radio frequency interference, and electrostatic discharge from various sources are the primary potential causes of these interferences. Furthermore, these interferences have impacted the development and progression of 5G technology. High-frequency electromagnetic (EM) waves saturated the atmosphere. To make cables, we subject polymers to extrusion. Highly populated areas, undersea tunnels, and underground tunnels have been established. There are high-voltage DC and AC cable systems to manage the increasing capacity (Li et al., 2021). Nevertheless, in densely populated areas where it is challenging or impractical

DOI: 10.1201/9781003527602-6

to install a network of overhead transmission, the transmission and distribution of power depend on underground cables (Liebscher et al., 2020). High-voltage, medium-voltage, and low-voltage applications use underground cables (Guo et al., 2020). Developing the next generation of cable insulation materials is crucial in order to enhance their electrical performance and elevate their working voltage levels (Guo et al., 2020).

This chapter will examine the most recent advancements in nanopolymer composites and discuss their applications in the electrical and electronic sectors.

6.2 LATEST ADVANCEMENTS IN POLYMER NANOCOMPOSITES IN THE ELECTRONIC SECTOR

6.2.1 Enhanced Electrical Conductivity

Enhancing the electrical conductivity of polymer nanocomposites (PNCs) is crucial, as they have potential applications in external fields that include mainly the development of hybrid nanofillers and surface functionalization.

Current studies have prioritized enhancing the electrical conductivity of PNC. Researchers have suggested several approaches (Li et al., 2021) to achieve superior electrical and thermal conductivities in polymer composites by adding fewer quantities of fillers. The thermodynamic and kinetic differences (like different melting points) between the nanofillers and the matrix can cause double percolation (Li et al., 2021). This happens when the nanofillers selectively localize in a matrix that doesn't mix with anything else.

This technique promotes a dominant arrangement of conductive fillers inside a single matrix phase. Because of this, randomly dispersed CPCs (R-CPCs) need a lot more filler to get to the same level of conductivity as double-percolated CPCs (Guo et al., 2020; Liebscher et al., 2020). An insulating polymer matrix and a conductive filler like carbon black (CB), carbon nanotubes (CNTs), or graphene make up conducting PNCs. This is why researchers are interested in them (Guo et al., 2016; Haggenmueller et al., 2007; Li et al., 2018a; Mutiso and Winey, 2015). A connected network of fillers is established when the number of fillers in the PNCs exceeds the percolation threshold, leading to a shift from nonconductivity to conductivity (Balberg, 1987).

6.2.2 Shielding of EMI

EMI represents a novel form of pollution arising from recent technological advancements and the extensive utilization of electronic devices. The increase in high-energy EM signals in the environment, along with EM coupling, frequency interference, radio electrostatic discharge, and EM conduction or induction from various sources, represents potential primary causes of these interferences (Wang et al., 2009). EMI adversely affects electrical systems and electronic devices, particularly in high-end applications such as communication, medical, military, and remote sensing (Mathur and Raman, 2020). EM radiation adversely affects the human body by disrupting microwave bands and high-frequency radio frequencies (RF) (Joo and Lee, 2000;

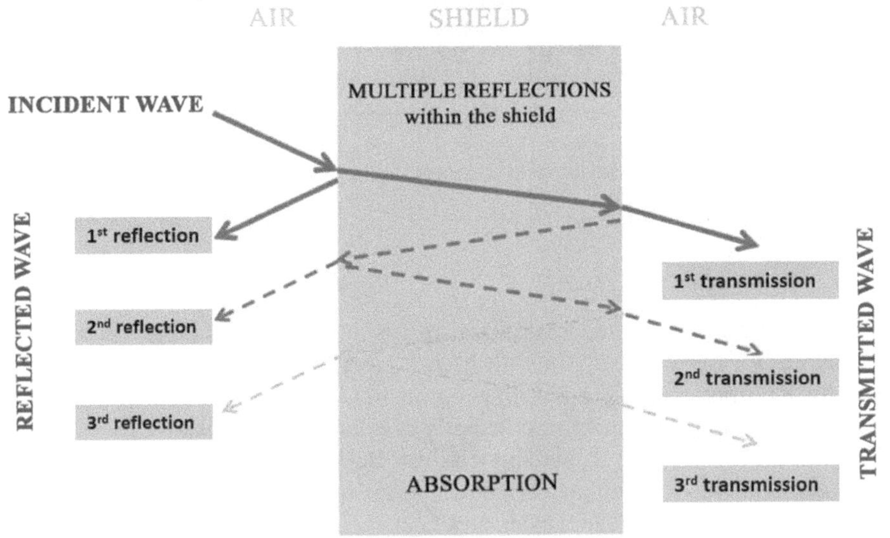

FIGURE 6.1 Schematic illustration of the electromagnetic shielding mechanism [100].

Kasevich, 2002). Ongoing exposure to EM radiation elevates the risk of heart disease, cancer, asthma, migraines, and miscarriages (Zhang et al., 2014). Due to the detrimental effects of EMI, researchers have focused on polymeric materials over the past few decades to effectively reduce EM radiation. Polymeric materials demonstrate enhanced EMI shielding performance relative to metal-based shielding materials due to their higher conductivity, superior mechanical properties, favorable permeability, inherent flexibility, and chemical resistance (Chung, 2000, 2001; Joo and Epstein, 1994; Rathi and Panwar, 2018; Schulz, 1983; Yang et al., 2005). Internally conducting polymer (ICP)-based EMI shielding materials exhibit significantly enhanced electrical conductivity and corrosion resistance (Håkansson et al., 2007; Jiang et al., 2019; Jing et al., 2005). The incorporation of suitable conducting fillers can enhance electrical conductivity. EMI shielding in polymeric materials is primarily achieved through absorption and reflection mechanisms (Colaneri and Schacklette, 1992; Wang and Jing, 2005) (Figure 6.1).

6.2.3 Dielectric Properties

The dielectric properties of PNCs have been extensively studied for use in capacitors and other electronic components. Researchers have developed materials with high dielectric constants and low dielectric loss by carefully selecting and dispersing nanoparticles like barium titanate or silica. These properties are essential for the development of miniaturized electronic devices with improved performance. Due to their superior power density and remarkable rate capabilities, high-tech electronics and electric power systems commonly use dielectrics as electrical energy storage materials (Azizi et al., 2017; Sarjeant et al., 1998). Because of its extremely high power density, quick charge/discharge speed, and extended cycling life, the dielectric capacitor is currently one of the most promising options when compared to other different types of energy storage devices such as rechargeable ion batteries like Mg, Li, and Zn ion batteries (Chi et al., 2019; Chu et al., 2006; Khanchaitit et al., 2013; Li et al., 2007, 2018b). Even though batteries have greater capacity for charging and discharging, their use is restricted in some locations due to environmental contamination caused by the chemical components of rechargeable ion batteries (Liu et al., 2018; Pang et al., 2018).

6.2.4 Thermal Management

Effective thermal management is critical in electronic devices to prevent overheating and ensure longevity. PNCs with high thermal conductivity have been developed by incorporating materials like boron nitride nanosheets or carbon-based fillers. These composites help in efficiently dissipating heat, making them suitable for use in high-power electronic applications. Bulk polymers, in contrast to fibers and films, often have irregularly oriented and infrequently aligned polymer chains, which makes thermal enhancement and heat transport difficult. Moreover, films and fibers with a high characteristic ratio offer increased surface area for effective heat transfer. Many techniques have been used to increase the thermal conductivity of bulk polymers in order to overcome their thermal constraints. The techniques used to increase the thermal conductivity of polymers are shown in Figure 6.2 (Ma et al., 2021). The most popular and well-studied methods include (i) creating 3D network topologies by adding one or more thermally conductive fillers to the polymer matrix, (ii) using mechanical stretching to align polymer chains, and (iii) creating hybrid composites made of two or more polymers, and (iv) to enhance the thermal conduction channels, polymer strands might be oriented differently. The planned uses and the processability of a particular polymer determine the technology to be used (Table 6.1).

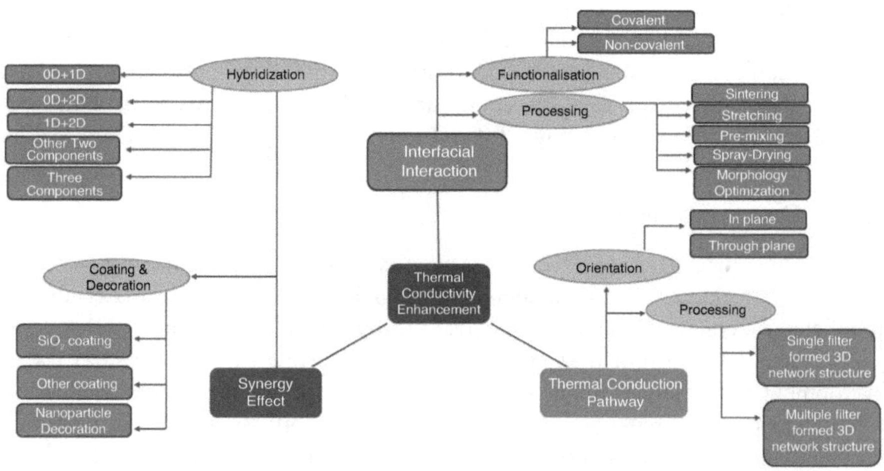

FIGURE 6.2 Techniques for improving polymers' thermal conductivity.

TABLE 6.1
Thermal Conductivity of Fillers

Material Type	Fillers	Thermal Conductivity ($W/m^{-1}K^{-1}$)	Reference
Metals	Ag	429	Zhou et al. (2017)
	Cu	400	Sun and Hobert (2021)
	Al	237	Li et al. (2020)
	Mg	156	Si et al. (2020)
Carbon-based	Carbon fibers (CFs)	530–1,100	Fugallo et al. (2014)
	Carbon nanotubes (CNTs)	600–4,000	Fugallo et al. (2014)
	Graphite	100–400	Fugallo et al. (2014)
	Graphene	1,500–5,000	Jo et al. (2013)
Ceramics	Boron nitride (BN)	250–360	Huang et al. (2011)
	Aluminum nitride (AlN)	150–26	Hirao et al. (2012)
	Silicon carbide (SiC)	490	Hong et al. (2017)
	Aluminum oxide (Al_2O_3)	30–42	Thostenson et al. (2001)

6.2.5 Mechanical Properties

The mechanical properties of PNCs have also improved significantly compared to traditional polymers. The addition of nanoparticles enhances the strength, flexibility, and toughness of the composites, which is beneficial for applications in flexible and wearable electronics. This enhancement ensures that the materials can withstand the mechanical stresses encountered during use. CNT and graphene-related materials have been the subject of extensive research due to their remarkable mechanical strength and low mass density. As a result, these materials have gained considerable attention as engineering materials suitable for a wide range of demanding applications (Saha, 2024).

6.2.6 Environmental Stability

Advancements have also been made in improving the environmental stability of PNCs. By incorporating nanoparticles that provide UV protection or improve thermal stability, researchers have developed materials that are more resistant to degradation under harsh environmental conditions. This stability is essential for the longevity and reliability of electronic devices used in outdoor or extreme environments. The use of green PNCs in various environmental preservation applications has garnered significant attention. Owing to their environmentally favorable composition, they find frequent applications in the automotive, construction, packaging, and healthcare sectors. Biodegradable polymers, derived from renewable sources such as bacteria, plants, and animals, offer environmentally responsible solutions to the issues posed by plastic waste. Green composite materials are made by combining natural polymers with plant fibers. Bacteria can degrade the fibers and resins used in green composites. Blending organic polymers with natural fillers produces green polymer nanocomposites with unique properties (Tajik et al., 2021).

6.2.7 Sensing Applications

Polymer nanocomposites have shown great promise in sensing applications due to their enhanced sensitivity and selectivity. Polymers are essential components of contemporary electronics. With countless innovative applications, they can be employed in their pure, composite, or nanocomposite forms for a variety of home and commercial uses. Significant progress in the development of nanomaterials-based electrochemical sensor platforms has aided in creating new technologies to protect human and environmental health. Because polymer nanocomposites have very good electrocatalytic activities, higher electrical conductivity, and an effective surface area compared to traditional polymers, researchers have recently increased their efforts to use them in the construction of electrochemical sensors (Abdel-Hakim and Sayed, 2024).

6.3 APPLICATIONS IN THE ELECTRONIC INDUSTRY

The design of cutting-edge devices for electrical and optoelectronic applications is closely linked to nanotechnology (Table 6.2). Polymer nanocomposites have found extensive applications in the electronics industry due to their unique properties and functionalities.

TABLE 6.2
Applications of Polymer Nanocomposites in the Electronic Industry

S.No	Materials Used	Applications	References
1	Conductive polymers are useful in sensor applications. These include polythiophene (PT), polypyrrole (PPY), poly (3,4-ethyl-oxy thiophene) (PEDOT), polyaniline (PANI), trans-polyacetylene (TPA), and poly (p-phenylene vinylene) (PPV)	Chemical sensors	John (2020)
2	Polymer nanocomposites (PNCs) are made up of nanofillers like metal nanoparticles, conductive polymers with graphene (Grp) or carbon nanotubes (CNTs), polyfluorene nanocomposites, or poly (phenylene)	Electroluminescent devices	Kim and Yang (2020)
3	Nanofillers such as metal nanoparticles, conductive polymers with graphene (Grp), carbon nanotubes (CNTs), polyfluorene (PF), and poly(phenylene vinylene) (PPV) nanocomposites make up polymer nanocomposites (PNCs)	Electro catalysis	Kim and Yang (2020)
4	There are different kinds of polymer nanocomposites (PNCs); some are conductive polymers with graphene, carbon nanotubes, metal nanoparticles, PVDF nanocomposites, and PEG nanocomposites	Development of batteries	Kim and Yang (2020)

(Continued)

TABLE 6.2 (*Continued*)
Applications of Polymer Nanocomposites in the Electronic Industry

S.No	Materials Used	Applications	References
5	Poly (vinyl alcohol) (PVA), polydimethylsiloxane (PDMS), poly (methyl methacrylate) (PMMA), and polyimide (PI) nanocomposites, thermochromic VO_2 nanocomposites, WO_3/ITO nanocomposites	Smart windows	Kausar (2021); Kumari et al. (2023) and Murillo-Quirós et al. (2022)
6	Various carbon-based nanoparticles such as carbon nanotubes, carbon black, graphene oxide, graphene nanoplatelets, graphene quantum dots, and nano-diamonds, metal oxide-based nanoparticles like Fe_3O_4 and TiO_2, cellulose-based nanoparticles like cellulose nanocrystals and nano-cellulose gel, and other nanomaterials like nano-clay, titanium nitride (TiN), gold nanorods (AuNRs), organic nanoparticles, silica, sepiolite, silsesquioxane, and hydroxyapatite nanofillers combine to form polyvinyl alcohol (PVA)	Memory devices	Bhanushali et al. (2022) and Ntanovasilis et al. (2021)
7	Poly(3-hexylthiophene) (P3HT) nanocomposites, poly(thieno[3,4-b]thiophene-co-benzodithiophene) (PTB7) nanocomposites, poly(3,4-ethylenedioxythiophene) (PEDOT) nanocomposites, nanocomposite coatings, such as TiO_2, ZnO, and CNT	Photovoltaic (PV) cells and photodiodes	Gayathri Mohan et al. (2024) Shah et al. (2022)
8	Polyaniline (PANI), polypyrrole (PPy), polythiophene (PTh), poly (3,4-ethylene dioxythiophene) (PEDOT), and polyethylene oxide (PEO) nanocomposites, polyimide (PI) nanocomposites, poly(vinyl alcohol) (PVA) nanocomposites	Supercapacitors	Islam et al. (2022), Sephra et al. (2021) and Yang (2012)
9	Organic semiconductors often use polymers such as polyaniline (PANI), polypyrrole (PPy), polythiophene (PTh), and poly(3,4-ethylene dioxythiophene) (PEDOT). Polymer/graphene nanocomposites are another type of material that combines polymers with graphene. Solution-processed organic semiconductors are commonly used, such as regioregular poly (3-hexylthiophene) (P3HT), poly (triarylamine), poly (3,3-dodecyl quaterthiophene) (PQT), poly (2,5-bis (3-tetradecyllthiophen-2-yl), and thieno [3,2-b] thiophene) (PBTTT). Additionally, polyaniline nanofibers and polyaniline/metal nanocomposites are materials of interest in the field.	Printable conductors, light-emitting diodes (LEDs), and field-effect transistors	Jiang et al. (2019), Nayak and Bilash Choudhary (2021), Pal et al. (2021) and Tong (2022)

(*Continued*)

TABLE 6.2 (*Continued*)
Applications of Polymer Nanocomposites in the Electronic Industry

S.No	Materials Used	Applications	References
10	Intrinsically conductive polymer-based nanocomposites, i.e., polyacetylene (Pan), polyaniline (PANI), polypyrrole (PPy) graphene-based polymer nanocomposites, i.e., polystyrene/graphene, poly(methyl methacrylate)/ graphene, polyethylene/graphene	Electromagnetic interference shielding	Kausar (2021) and Naghdi et al. (2018)
11	Polythiophene (PTh), polyaniline (PANI), polypyrrole (PPY), polystyrene (PS), polypropylene (PP), and poly(ethylene) (PE) are the compounds in this group of chemicals.	A variety of optoelectronic devices, such as liquid crystal displays (LCDs), touch screens, solar cells, organic light-emitting diodes (OLEDs), and electrochromic devices, use transparent conductive coatings.	Hassan et al. (2021) and Shah et al. (2022)
12	Polyaniline (PANI), polypyrrole (PPy), polythiophene (PTh), poly(3,4-ethylenedioxythiophene) (PEDOT), poly(para-phenylene) (PPP), polyfuran (PF), and poly(phenylene vinylene) (PPV) are the compounds in this group of chemicals.	Electrostatic dissipation, supercapacitors	Boruah et al. (2024), Pande et al. (2023) and Yang (2012)
13	Polypyrrole(PPy), polyaniline (PANI), polyacetylene, polyfuran, and polythiophene	Electromechanical actuators and various electrode applications	Maksimkin et al. (2022)

6.4 APPLICATIONS IN THE ELECTRICAL INDUSTRY

6.4.1 Breakdown Strength of an AC (Alternating Current) System

Polymer nanocomposites can serve as dielectric materials to strengthen underground cables, depending on the number and distribution of nanoparticles in the polymer host matrix. Researchers have documented several empirical investigations on the electrical breakdown resistance of nanocomposites (Gouda and Haiba, 2019; Li et al., 2015) (Table 6.3). Huang et al. (2011) showed that adding SiO_2 nanofillers that had been treated with octasilane improved LDPE's AC breakdown strength. Functionalized fumed SiO_2 nanofillers slightly increased XLPE's AC breakdown

TABLE 6.3
The Recent Development of AC Breakdown Strength of Polymer Composites for Power Cable Applications

Polymer Nanocomposites	Mean Size (nm)	Best Concentrations (wt%)	Key Findings	References
HDPE/$CaCO_3$	6–40	2	The mixture of HDPE and 1% g-maleic anhydride has a 21% higher breakdown strength than the pure HDPE sample.	Ali et al. (2021)
LDPE/Al_2O_3	30 ± 5	0.5	Breakdown strength is 35% higher than the pure LDPE sample.	Wang et al. (2016b)
PVC/SiO_2	6	5	Breakdown strength is 7% higher than the pure PVC sample.	Habashy et al. (2021)
HDPE/clay and LDPE/clay	10	1	HDPE/clay has 128% and LDPE/clay has 123.5% higher breakdown strength than the pure sample.	Mohamed (2015)
LDPE/Al_2O_3	45	3	Breakdown strength is 13.5% higher than the LDPE sample.	Chen et al. (2016)
PVC/SiO_2	6-	0.5	The treated SiO_2 sample enhances the breakdown strength by 14.1%, while the untreated sample enhances it by 7.7%.	Abdel-Gawad et al. (2018)

strength (Abdel-Gawad et al., 2017; Tanaka et al., 2011). When polyhedral oligomeric silsesquioxane was added to LDPE, it made it 9% stronger against TiO_2 nanoparticles and 7% stronger against SiO_2 nanoparticles. The addition of vinyl silane to the TiO_2 nanoparticle surface increased the AC breakdown strength of the PVC/TiO_2 nanocomposites by 11% (Abdel-Gawad et al., 2017; Zazoum et al., 2016).

6.4.2 Breakdown Strength of a DC System

Table 6.4 provides a summary of the key findings on the DC breakdown strength of polymer nanocomposites that have been reported in the literature. Significant variations exist in the findings of many authors regarding nanocomposites that are supposedly identical, such as LDPE/Al_2O_3 with a nanoparticle weight fraction of 0.5% as described in references (Wang et al., 2016b) and LDPE/MgO with a

TABLE 6.4
The Recent Development of DC Breakdown Strength of Polymer Composites for Power Cable Applications

Polymer Nanocomposites	Avg. Size (nm)	Optimum Concentrations	Key Findings	Ref.
XLPE/GO	50	0.01% by weight	Functionalized graphene oxide quantum wells improve the breakdown strength by 33.3% over XLPEs.	Ali et al. (2021)
PVC/ZnO	28%–41.5%	0.5% by weight	The breakdown strength exhibits a 45% increase as compared to the pure sample.	Mansour et al. (2016)
LDPE/Al_2O_3	10–6.	0.5% weight fraction	The breakdown strength shows a 35% improvement as compared to the pure sample.	Wang et al. (2016b)
PVC/$CaCO_3$	10–30.	5% weight fraction	The breakdown strength exhibits a 59.5% increase as compared to the original sample.	Sugumaran (2015)
PVC/SiO_2	10–30.	5% by weight	The breakdown strength exhibits a 48.8% increase as compared to the original sample.	Sugumaran (2015)

nanoparticle weight fraction of 5% as published in references (Ahmed Dabbak et al., 2018; Sugumaran, 2015). The significant need for DC breakdown strength on space charge characteristics, influenced by manufacturing techniques and material purity, accounts for the discrepancies (Zhang et al., 2014). When SiO_2 nanoparticles were mixed with PE, the DC breakdown strength went down. This happened most noticeably when the nanoparticles' weight percentage was between 1% and 10%. This behavior is similar to the results reported concerning the AC breakdown strength. However, the addition of SiO_2 nanoparticles to PVC greatly increased the DC breakdown strength (Sugumaran, 2015).

6.4.3 Impulse Breakdown Strength in Power Cables

Impulse breakdown strength refers to a material's ability to withstand high-voltage impulses without experiencing an electrical breakdown. There is a limited number of studies on the impulse breakdown strength of nanocomposites (Murata et al., 2005). Table 6.5 concisely summarizes the main findings acquired from the literature and demonstrates a decrease in impulse breakdown strength. As the concentration of nanoparticles increases beyond around 1%, except for the findings from (Okuzumi et al., 2008), there is a noticeable effect. The percentage gains are typically not as dramatic as those seen under AC and DC (Wang et al., 2012).

TABLE 6.5
Recent Advances in Polymer Composite (PCs) Dielectrics for High-Temperature Applications

Polymer Nanocomposites	Average Size (nm)	Optimum Concentrations	Key Findings	Reference
LDPE/MgO	10	2 wt%	Break down strength enhanced with 21%	Wang et al. (2012)
LDPE/MgO	50	1 wt%	Break down strength enhanced with 4%	Wang et al. (2016a)
LDPE/SiO2	30	1 wt%	Break down strength enhanced with 10%	Hoang et al. (2016)
LDPE/Al_2O_3	<100	0.5 wt%	Break down strength enhanced with 30%	Liu et al. (2017)

6.4.4 Permittivity, Dielectric, and Conductivity Loss of Nanodielectrics in Power Cables

Wang et al. (2015) found that an LDPE/Al_2O_3 nanocomposite had a lower permittivity than pure LDPE. The researchers came to the conclusion that the lower permittivity was because of functionalized Al_2O_3 nanoparticles, which made it harder for charge carriers to move. The permittivity exhibited frequency independence throughout, covering 100 Hz to 1 MHz. The presence of charge carriers at the electrode and LDPE contacts suggests occlusion of the electrode (Abdel-Gawad et al., 2018). The study focuses on nanocomposites consisting of PVC and functionalized TiO_2. The permittivity was reduced by approximately 43% in comparison to pure PVC and by roughly 33% in comparison to polyvinyl chloride/unaltered TiO_2 nanocomposites. The size and type of the NPs, as well as the surface modification coupling agent, have an impact on permittivity. Table 6.4 provides a thorough overview of the primary discoveries regarding the dielectric properties of polymer nanocomposites. The study investigates electrical permittivity, electrical conductivity, and dielectric loss.

The permittivity remained constant while the frequency ranged from 100 Hz to 1 MHz. This case demonstrated electrode polarization due to the difficulty of charge carriers in traversing the interfaces between the electrode and LDPE. Scientists observed a comparable trend in PVC/functionalized TiO_2 nanocomposites (Abdel-Gawad et al., 2018). The permittivity decreased by roughly 43% compared to pure PVC and approximately 33% compared to PVC/unmodified TiO_2 nanocomposites. The size and type of the NPs, as well as the surface modification coupling agent, have an impact on permittivity. As the weight fraction of Al_2O_3 increased, the electrical conductivity of LDPE/Al_2O_3 NCs rose significantly. At a loading of 0.5 wt%, the rise nearly multiplied by 10 but became restricted at higher loadings. Nevertheless, adding 3% uncoated MgO and Al_2O_3 nanoparticles to LDPE nanocomposites made them much less conductive. LDPE nanocomposites' conductivity decreased when Al_2O_3 nanoparticles coated with silanes had terminal alkyl groups of different lengths added, even though the concentration was only 1% by weight remained unchanged (Liu et al., 2017).

6.4.5 Dielectric Capacitors for Electrical Applications

Dielectric capacitors are preferable for electrical energy conservation because of their fast charge/discharge rate, exceptionally high storage options with high energy density, and extended lifespan, surpassing alternatives such as rechargeable ion batteries (Dastan et al., 2015).

Li Q et al. proposed the incorporation of ceramic filler into a polymer matrix in order to create a dielectric polymer composite (Rajib et al., 2016). This would address the issue of poor energy density under elevated temperatures. Some examples of gate dielectrics with low dielectric constants are poly-4-vinyl phenol, polymethyl methacrylate, and polyvinyl alcohol. The incorporation of inorganic fillers such as HfO and TiO_2 into the polymer matrix significantly increased the dielectric constant while maintaining a very low dielectric loss (Dastan et al., 2015). However, the success of incorporating nanoparticles into the polymer network depends on their compatibility and dispersion. Multiple inorganic fillers can be added to a polymer matrix to improve its electrical stability and decrease space charge accumulation. Recent studies on dielectrics made from polymer nanocomposites have substantiated this. We can enhance the energy density in polymer nanocomposite dielectrics (Hu et al., 2020) by adjusting the quantity of ceramic filler and modifying the surface, thereby achieving the desired breakdown strength and dielectric characteristic. Hu et al. incorporated h-BN or hexagonal boron nitride into the polymer matrix to form a composite material with dielectric properties. Incorporating h-BN into the polymer matrix effectively inhibits electrical conductivity. As a result, these polymer composites exhibit a significant energy density and exceptional discharging capabilities, particularly under elevated temperatures (Hu et al., 2020). The researchers have successfully created polymer-based composite dielectrics that demonstrate a remarkable energy density of up to 6.0 J/cm^3 at a temperature of 16°C and when subjected to an electric field intensity of 60 MV/m. Rajib et al.'s research facilitated the successful accomplishment of BaTiO by incorporating it into a polyimide (PI) matrix. It is crucial to be able to find a high energy density at high temperatures while simultaneously applying a powerful electric field and ensuring efficient charge–discharge performance (Rajib et al., 2016).

6.4.6 Improvements in the Microstructure of Nanofiller-Based Nano Polymer Composites for Electrical Applications

The researchers discovered that variations in applied electric fields and temperature significantly affect the nanofiller's morphology, microstructure, alignment, and quantity. Moreover, the microstructure of nanocomposites has a substantial impact on the breakdown strength of polymer composite dielectrics. The phase-field model study shows that nanocomposites with parallel nanosheets are better at withstanding higher electric fields than those with nanofibers inserted vertically.

6.5 CONCLUSION AND FUTURE OUTLOOK

Because polymer nanocomposites have improved qualities over traditional polymers, they have become disruptive in the electrical and electronic industries. The use of nanoscale fillers, including metal nanoparticles, CNTs, graphene, and nano-clays, has greatly enhanced the mechanical, electrical, and thermal properties of polymer matrices. These developments have made it possible to create high-performance materials that are necessary for sensors, flexible electronics, and EMI shielding, among other uses. Additionally, researchers have added numerous strategies for optimizing the polymer nanocomposite for diverse industries. Production of EMI shielding and power cable-based polymer nanocomposite materials is also necessary for future uses. One of the key problems that future technology will need to solve is the EMI. Experts predict the market will expand at a compound annual growth rate of 19.50% to reach USD 44.2 billion by 632.

REFERENCES

Abdel-Gawad, N.M.K., El Dein, A.Z., Mansour, D.-E.A., Ahmed, H.M., Darwish, M.M.F., Lehtonen, M., 2018. Multiple enhancement of PVC cable insulation using functionalized SiO 2 nanoparticles based nanocomposites. *Electr. Power Syst. Res.* 163, 612–625. https://doi.org/10.1016/j.epsr.2017.11.011

Abdel-Gawad, N.M.K., El Dein, A.Z., Mansour, D.-E.A., Ahmed, H.M., Darwish, M.M.F., Lehtonen, M., 2017. Enhancement of dielectric and mechanical properties of polyvinyl chloride nanocomposites using functionalized TiO_2 nanoparticles. *IEEE Trans. Dielect. Electr. Insul.* 24, 3490–3499. https://doi.org/10.1109/TDEI.2017.006692

Abdel-Hakim, M., Sayed, M.M., 2024. Conducting polymers sensor, in: Ali, G.A.M., Chong, K.F., Makhlouf, A.S.H. (Eds.), *Handbook of Nanosensors*. Springer Nature, Cham, pp. 1–34. https://doi.org/10.1007/978-3-031-16338-8_71-1

Ahmed Dabbak, S.Z., Illias, H.A., Ang, B.C., Abdul Latiff, N.A., Makmud, M.Z.H., 2018. Electrical properties of polyethylene/polypropylene compounds for high-voltage insulation. *Energies* 11, 1448. https://doi.org/10.3390/en11061448

Ali, S.F.A., Elsad, R.A., Mansour, S.A., 2021. Enhancing the dielectric properties of compatibilized high-density polyethylene/calcium carbonate nanocomposites using high-density polyethylene-g-maleic anhydride. *Polym. Bull.* 78, 1393–1405. https://doi.org/10.1007/s00289-020-03164-w

Azizi, A., Gadinski, M.R., Li, Q., AlSaud, M.A., Wang, J., Wang, Y., Wang, B., Liu, F., Chen, L., Alem, N., Wang, Q., 2017. High-performance polymers sandwiched with chemical vapor deposited hexagonal boron nitrides as scalable high-temperature dielectric materials. *Adv. Mater.* 29, 1701864. https://doi.org/10.1002/adma.201701864

Balberg, I., 1987. Recent developments in continuum percolation. *Phil. Mag. B* 56, 991–1003. https://doi.org/10.1080/13642818708215336

Bhanushali, H., Amrutkar, S., Mestry, S., Mhaske, S.T., 2022. Shape memory polymer nanocomposite: a review on structure–property relationship. *Polym. Bull.* 79, 3437–3493. https://doi.org/10.1007/s00289-021-03686-x

Boruah, B., Borah, S., Deka, M., 2024. Electroactive polymer-based nanostructures and nanocomposites for sensing applications, in: Mohanta, D., Chakraborty, P. (Eds.), *Nanoscale Matter and Principles for Sensing and Labeling Applications, Advanced Structured Materials*. Springer Nature Singapore, Singapore, pp. 243–265. https://doi.org/10.1007/978-981-99-7848-9_12

Chen, X., Murdany, D., Liu, D., Andersson, M., Gubanski, S.M., Gedde, U.W., Suwarno, 2016. AC and DC pre-stressed electrical trees in LDPE and its aluminum oxide nanocomposites. *IEEE Trans. Dielect. Electr. Insul.* 23, 1506–1514. https://doi.org/10.1109/TDEI.2016.005622

Chi, Q., Gao, Z., Zhang, T., Zhang, C., Zhang, Y., Chen, Q., Wang, X., Lei, Q., 2019. Excellent energy storage properties with high-temperature stability in sandwich-structured polyimide-based composite films. *ACS Sustain. Chem. Eng.* 7, 748–757. https://doi.org/10.1021/acssuschemeng.8b04370

Chu, B., Zhou, X., Ren, K., Neese, B., Lin, M., Wang, Q., Bauer, F., Zhang, Q.M., 2006. A dielectric polymer with high electric energy density and fast discharge speed. *Science* 313, 334–336. https://doi.org/10.1126/science.1127798

Chung, D.D.L., 2001. Electromagnetic interference shielding effectiveness of carbon materials. *Carbon* 39, 279–285. https://doi.org/10.1016/S0008-6223(00)00184-6

Chung, D.D.L., 2000. Materials for electromagnetic interference shielding. *J. Mater. Eng. Perform.* 9, 350–354. https://doi.org/10.1361/105994900770346042

Colaneri, N.F., Schacklette, L.W., 1992. EMI shielding measurements of conductive polymer blends. *IEEE Trans. Instrum. Meas.* 41, 291–297. https://doi.org/10.1109/19.137363

Dastan, D., Gosavi, S.W., Chaure, N.B., 2015. Studies on electrical properties of hybrid polymeric gate dielectrics for field effect transistors. *Macromol. Symp.* 347, 81–86. https://doi.org/10.1002/masy.201400042

Fugallo, G., Cepellotti, A., Paulatto, L., Lazzeri, M., Marzari, N., Mauri, F., 2014. Thermal conductivity of graphene and graphite: collective excitations and mean free paths. *Nano Lett.* 14, 6109–6114. https://doi.org/10.1021/nl502059f

Gayathri Mohan, G.M., George, J., Balachandran, M., 2024. Polymer-nanocarbon composites: a promising strategy for enhanced performance of organic solar cells. *Emerg. Mater.* 7, 17–33. https://doi.org/10.1007/s42247-023-00604-x

Gouda, O.E., Haiba, A.S., 2019. Measurements of dielectric properties of high density polyethylene-nano-composites. *Measurement* 134, 624–633. https://doi.org/10.1016/j.measurement.2018.10.029

Guo, B., Tang, Z., Zhang, L., 2016. Transport performance in novel elastomer nanocomposites: mechanism, design and control. *Prog. Polym. Sci.* 61, 29–66. https://doi.org/10.1016/j.progpolymsci.2016.06.001

Guo, M., Kashfipour, M.A., Li, Y., Dent, R.S., Zhu, J., Maia, J.M., 2020. Structure-rheology-Property relationships in double-percolated polypropylene/poly(methyl methacrylate)/boron nitride polymer composites. *Compos. Sci. Technol.* 198, 108306. https://doi.org/10.1016/j.compscitech.2020.108306

Habashy, M.M., Abd-Elhady, A.M., Elsad, R.A., Izzularab, M.A., 2021. Performance of PVC/SiO2 nanocomposites under thermal ageing. *Appl. Nanosci.* 11, 2143–2151. https://doi.org/10.1007/s13204-018-00941-y

Haggenmueller, R., Guthy, C., Lukes, J.R., Fischer, J.E., Winey, K.I., 2007. Single wall carbon nanotube/polyethylene nanocomposites: thermal and electrical conductivity. *Macromolecules* 40, 2417–2421. https://doi.org/10.1021/ma0615046

Håkansson, E., Amiet, A., Nahavandi, S., Kaynak, A., 2007. Electromagnetic interference shielding and radiation absorption in thin polypyrrole films. *Eur. Polym. J.* 43, 205–213. https://doi.org/10.1016/j.eurpolymj.2006.10.001

Hassan, T., Salam, A., Khan, A., Khan, S.U., Khanzada, H., Wasim, M., Khan, M.Q., Kim, I.S., 2021. Functional nanocomposites and their potential applications: a review. *J. Polym. Res.* 28, 36. https://doi.org/10.1007/s10965-021-02408-1

Hirao, K., Zhou, Y., Hyuga, H., Ohji, T., Kusano, D., 2012. High thermal conductivity silicon nitride ceramics. *J. Korean Ceram. Soc.* 49, 380–384. https://doi.org/10.4191/kcers.2012.49.4.380

Hoang, A., Pallon, L., Liu, D., Serdyuk, Y., Gubanski, S., Gedde, U., 2016. Charge transport in LDPE nanocomposites part I—experimental approach. *Polymers* 8, 87. https://doi.org/10.3390/polym8030087

Hong, H., Kim, J., Kim, T., 2017. Effective assembly of nano-ceramic materials for high and anisotropic thermal conductivity in a polymer composite. *Polymers* 9, 413. https://doi.org/10.3390/polym9090413

Hu, H., Zhang, F., Luo, S., Chang, W., Yue, J., Wang, C.-H., 2020. Recent advances in rational design of polymer nanocomposite dielectrics for energy storage. *Nano Energy* 74, 104844. https://doi.org/10.1016/j.nanoen.2020.104844

Huang, X., Jiang, P., Tanaka, T., 2011. A review of dielectric polymer composites with high thermal conductivity. *IEEE Electr. Insul. Mag.* 27, 8–16. https://doi.org/10.1109/MEI.2011.5954064

Islam, Md. Mahinur, Sheikh, Md .S.I., Susan, Md .A.B.H., Islam, Md. Mominul, 2022. Conjugated polymers as the materials for supercapacitor electrodes, in: Gupta, R. K. (Ed.), *Organic Electrodes, Engineering Materials*. Springer International Publishing, Cham, pp. 265–287. https://doi.org/10.1007/978-3-030-98021-4_15

Jiang, D., Murugadoss, V., Wang, Y., Lin, J., Ding, T., Wang, Z., Shao, Q., Wang, C., Liu, H., Lu, N., Wei, R., Subramania, A., Guo, Z., 2019. Electromagnetic interference shielding polymers and nanocomposites - a review. *Poly. Rev.* 59, 280–337. https://doi.org/10.1080/15583724.2018.1546737

Jing, X., Wang, Y., Zhang, B., 2005. Electrical conductivity and electromagnetic interference shielding of polyaniline/polyacrylate composite coatings. *J. Appl. Poly. Sci.* 98, 2149–2156. https://doi.org/10.1002/app.22387

Jo, I., Pettes, M.T., Kim, J., Watanabe, K., Taniguchi, T., Yao, Z., Shi, L., 2013. Thermal conductivity and phonon transport in suspended few-layer hexagonal boron nitride. *Nano. Lett.* 13, 550–554. https://doi.org/10.1021/nl304060g

John, B., 2020. Polymer nanocomposite-based electrochemical sensors and biosensors, in: Sasani Ghamsari, M., Dhara, S. (Eds.), *Nanorods and Nanocomposites*. IntechOpen. https://doi.org/10.5772/intechopen.86826

Joo, J., Epstein, A.J., 1994. Electromagnetic radiation shielding by intrinsically conducting polymers. *Appl. Phys. Lett.* 65, 2278–2280. https://doi.org/10.1063/1.112717

Joo, J., Lee, C.Y., 2000. High frequency electromagnetic interference shielding response of mixtures and multilayer films based on conducting polymers. *J. Appl. Phys.* 88, 513–518. https://doi.org/10.1063/1.373688

Kasevich, R.S., 2002. Cellphones, radars, and health [Speakout]. *IEEE Spectr.* 39, 15–16. https://doi.org/10.1109/MSPEC.2002.1021945

Kausar, A., 2021. Progress in green nanocomposites for high-performance applications. *Mater. Res. Innov.* 25, 53–65. https://doi.org/10.1080/14328917.2020.1728489

Khanchaitit, P., Han, K., Gadinski, M.R., Li, Q., Wang, Q., 2013. Ferroelectric polymer networks with high energy density and improved discharged efficiency for dielectric energy storage. *Nat. Commun.* 4, 2845. https://doi.org/10.1038/ncomms3845

Kim, H., Yang, S., 2020. Responsive smart windows from nanoparticle–polymer composites. *Adv. Funct. Mater.* 30, 1902597. https://doi.org/10.1002/adfm.201902597

Kumari, R., Gellanki, J., Kundale, S.S., Ustad, R.E., Dongale, T.D., Fu, Y., Pettersson, H., Kumar, S., 2023. Artificial synaptic characteristics of PVA:ZnO nanocomposite memristive devices. *APL Mater.* 11, 101124. https://doi.org/10.1063/5.0165205

Li, F., Liu, F., Liu, J., Gao, Y., Lu, Y., Chen, J., Yang, H., Zhang, L., 2018a. Thermo-mechanical coupling analysis of transient temperature and rolling resistance for solid rubber tire: numerical simulation and experimental verification. *Compos. Sci. Technol.* 167, 404–410. https://doi.org/10.1016/j.compscitech.2018.08.034

Li, J.Y., Zhang, L., Ducharme, S., 2007. Electric energy density of dielectric nanocomposites. *Appl. Phys. Lett.* 90, 132901. https://doi.org/10.1063/1.2716847

Li, Q., Yao, F.-Z., Liu, Y., Zhang, G., Wang, H., Wang, Q., 2018b. High-temperature dielectric materials for electrical energy storage. *Annu. Rev. Mater. Res.* 48, 219–243. https://doi.org/10.1146/annurev-matsci-070317-124435

Li, S., Yang, X., Hou, J., Du, W., 2020. A review on thermal conductivity of magnesium and its alloys. *J. Magnes. Alloys* 8, 78–90. https://doi.org/10.1016/j.jma.2019.08.002

Li, T., Wang, Y., Wang, Y., Sun, F., Xu, J., Lou, C., Lin, J., 2021. Preparation of flexible, highly conductive polymer composite films based on double percolation structures and synergistic dispersion effect. *Polym. Compos.* 42, 5159–5167. https://doi.org/10.1002/pc.26213

Li, W., Hillborg, H., Gedde, U.W., 2015. Influence of process conditions and particle dispersion on the ac breakdown strength of polyethylene-aluminium oxide nanocomposites. *IEEE Trans. Dielect. Electr. Insul.* 22, 3536–3542. https://doi.org/10.1109/TDEI.2015.005366

Liebscher, M., Domurath, J., Saphiannikova, M., Müller, M.T., Heinrich, G., Pötschke, P., 2020. Dispersion of graphite nanoplates in melt mixed PC/SAN polymer blends and its influence on rheological and electrical properties. *Polymer* 200, 122577. https://doi.org/10.1016/j.polymer.2020.122577

Liu, D., Hoang, A.T., Pourrahimi, A.M., Pallon, L.K.H., Nilsson, F., Gubanski, S.M., Olsson, R.T., Hedenqvist, M.S., Gedde, U.W., 2017. Influence of nanoparticle surface coating on electrical conductivity of LDPE/Al2O3 nanocomposites for HVDC cable insulations. *IEEE Trans. Dielect. Electr. Insul.* 24, 1396–1404. https://doi.org/10.1109/TDEI.2017.006310

Liu, G., Chi, Q., Zhang, Y., Chen, Q., Zhang, C., Zhu, K., Cao, D., 2018. Superior high rate capability of $MgMn_2$ O_4 /rGO nanocomposites as cathode materials for aqueous rechargeable magnesium ion batteries. *Chem. Commun.* 54, 9474–9477. https://doi.org/10.1039/C8CC05366A

Ma, H., Gao, B., Wang, M., Yuan, Z., Shen, J., Zhao, J., Feng, Y., 2021. Strategies for enhancing thermal conductivity of polymer-based thermal interface materials: a review. *J. Mater. Sci.* 56, 1064–1086. https://doi.org/10.1007/s10853-020-05279-x

Maksimkin, A.V., Dayyoub, T., Telyshev, D.V., Gerasimenko, A.Y., 2022. Electroactive polymer-based composites for artificial muscle-like actuators: a review. *Nanomaterials* 12, 2272. https://doi.org/10.3390/nano12132272

Mansour, S.A., Elsad, R.A., Izzularab, M.A., 2016. Dielectric properties enhancement of PVC nanodielectrics based on synthesized ZnO nanoparticles. *J. Polym. Res.* 23, 85. https://doi.org/10.1007/s10965-016-0978-5

Mathur, P., Raman, S., 2020. Electromagnetic interference (EMI): measurement and reduction techniques. *J. Electron. Mater.* 49, 2975–2998. https://doi.org/10.1007/s11664-020-07979-1

Mohamed, A.T., 2015. Experimental enhancement for dielectric strength of polyethylene insulation materials using cost-fewer nanoparticles. *Int. J. Electr. Power Energy Syst.* 64, 469–475. https://doi.org/10.1016/j.ijepes.2014.06.075

Murata, Y., Sekiguchi, Y., Inoue, Y., Kanaoka, M., 2005. Investigation of electrical phenomena of inorganic-filler/LDPE nanocomposite material, in: *Proceedings of 2005 International Symposium on Electrical Insulating Materials, 2005. (ISEIM 2005). Presented at the Proceedings of 2005 International Symposium on Electrical Insulating Materials, 2005. (ISEIM 2005),* IEEE, Kitakyushu, Japan, pp. 650–653 Vol. 3. https://doi.org/10.1109/ISEIM.2005.193453

Murillo-Quirós, N., Vega-Garita, V., Carmona-Calvo, A., Rojas-González, E.A., Starbird-Perez, R., Avendaño-Soto, E., 2022. Toward thermochromic VO2 nanoparticles polymer films based smart windows designed for tropical climates. *Polymers* 14, 4250. https://doi.org/10.3390/polym14194250

Mutiso, R.M., Winey, K.I., 2015. Electrical properties of polymer nanocomposites containing rod-like nanofillers. *Prog. Polym. Sci.* 40, 63–84. https://doi.org/10.1016/j.progpolymsci.2014.06.002

Naghdi, S., Rhee, K.Y., Hui, D., Park, S.J., 2018. A review of conductive metal nanomaterials as conductive, transparent, and flexible coatings, thin films, and conductive fillers: different deposition methods and applications. *Coatings* 8, 278. https://doi.org/10.3390/coatings8080278

Nayak, D., Bilash Choudhary, R., 2021. Conducting polymer-based emissive layer on efficiency of OLEDs, in: *Light-Emitting Diodes and Photodetectors – Advances and Future Directions [Working Title]*. IntechOpen. https://doi.org/10.5772/intechopen.98652

Ntanovasilis, G., Zaverdas, I., Ahmed, T., Markoulidis, F., Lekakou, C., 2021. Polymer blends and polymer nanocomposites for photovoltaic (PV) cells and an investigation of the material deposition techniques in PV cell fabrication. *J. Compos. Sci.* 5, 263. https://doi.org/10.3390/jcs5100263

Okuzumi, S., Murakami, Y., Nagao, M., Sekiguchi, Y., Reddy, C.C., Murata, Y., 2008. DC breakdown strength and conduction current of MgO/LDPE composite influenced by filler size, in: *2008 Annual Report Conference on Electrical Insulation and Dielectric Phenomena. Presented at the 2008 Annual Report Conference on Electrical Insulation and Dielectric Phenomena (CEIDP)*, IEEE, Quebec City, QC, Canada, pp. 722–725. https://doi.org/10.1109/CEIDP.2008.4772933

Pal, K., Si, A., Stephen, R., Thomas, S., 2021. Conductive polymer nanocomposites for organic light-emitting diodes (OLEDs), in: Hussain, C. M., Thomas, S. (Eds.), *Handbook of Polymer and Ceramic Nanotechnology*. Springer International Publishing, Cham, pp. 281–292. https://doi.org/10.1007/978-3-030-40513-7_14

Pande, S., Pandit, B., Shaikh, S.F., Ubaidullah, M., 2023. Conductive polymer and composites for supercapacitor applications, in: Gupta, R. K. (Ed.), *Recent Advancements in Polymeric Materials for Electrochemical Energy Storage, Green Energy and Technology*. Springer Nature Singapore, Singapore, pp. 71–92. https://doi.org/10.1007/978-981-99-4193-3_5

Pang, Q., Sun, C., Yu, Y., Zhao, K., Zhang, Z., Voyles, P.M., Chen, G., Wei, Y., Wang, X., 2018. $H_2 V_3 O_8$ nanowire/graphene electrodes for aqueous rechargeable zinc ion batteries with high rate capability and large capacity. *Adv. Energy Mater.* 8, 1800144. https://doi.org/10.1002/aenm.201800144

Rajib, M., Martinez, R., Shuvo, M., Karim, H., Delfin, D., Afrin, S., Rodriguez, G., Chintalapalle, R., Lin, Y., 2016. Enhanced energy storage of dielectric nanocomposites at elevated temperatures. *Int. J. Appl. Ceramic Technol.* 13, 125–132. https://doi.org/10.1111/ijac.12410

Rathi, V., Panwar, V., 2018. Electromagnetic interference shielding analysis of conducting composites in near- and far-field region. *IEEE Trans. Electromagn. Compat.* 60, 1795–1801. https://doi.org/10.1109/TEMC.2017.2780883

Saha, A., 2024. Polymer nanocomposites: a review on recent advances in the field ofgreen polymer nanocomposites. *CNANO* 20, 706–716. https://doi.org/10.2174/0115734137274950231113050300

Sarjeant, W.J., Zirnheld, J., MacDougall, F.W., 1998. Capacitors. *IEEE Trans. Plasma Sci.* 26, 1368–1392. https://doi.org/10.1109/27.736020

Schulz, R.A., 1983. A frequency response method for determining the parameters of high-performance DC motors. *IEEE Trans. Ind. Electron.* IE-30, 39–42. https://doi.org/10.1109/TIE.1983.356701

Sephra, P.J., Baraneedharan, P., Arulraj, A., 2021. Nanoelectronics devices (field-effect transistors, electrochromic devices, light-emitting diodes, dielectrics, neurotransmitters), in: Shahabuddin, S., Pandey, A. K., Khalid, M., Jagadish, P. (Eds.), *Advances in*

Hybrid Conducting Polymer Technology, Engineering Materials. Springer International Publishing, Cham, pp. 77–100. https://doi.org/10.1007/978-3-030-62090-5_4

Shah, S.S., Aziz, M.A., Mahfoz, W., Al-Betar, A.-R., 2022. Conducting polymers based nanocomposites for supercapacitors, in: Thomas, S., Gueye, A. B., Gupta, R. K. (Eds.), *Nanostructured Materials for Supercapacitors, Advances in Material Research and Technology*. Springer International Publishing, Cham, pp. 485–511. https://doi.org/10.1007/978-3-030-99302-3_22

Si, W., Sun, J., He, X., Huang, Y., Zhuang, J., Zhang, J., Murugadoss, V., Fan, J., Wu, D., Guo, Z., 2020. Enhancing thermal conductivity *via* conductive network conversion from high to low thermal dissipation in polydimethylsiloxane composites. *J. Mater. Chem.* C8, 3463–3475. https://doi.org/10.1039/C9TC06968B

Sugumaran, C.P., 2015. Experimental study on dielectric and mechanical properties of PVC cable insulation with SiO2/CaCO3 nanofillers, in: *2015 IEEE Conference on Electrical Insulation and Dielectric Phenomena (CEIDP). Presented at the 2015 IEEE Conference on Electrical Insulation and Dielectric Phenomena - (CEIDP)*, IEEE, Ann Arbor, MI, USA, pp. 503–506. https://doi.org/10.1109/CEIDP.2015.7352072

Sun, H., Hobert, O., 2021. Temporal transitions in the post-mitotic nervous system of Caenorhabditis elegans. *Nature* 600, 93–99. https://doi.org/10.1038/s41586-021-04071-4

Tajik, S., Beitollahi, H., Nejad, F.G., Dourandish, Z., Khalilzadeh, M.A., Jang, H.W., Venditti, R.A., Varma, R.S., Shokouhimehr, M., 2021. Recent developments in polymer nanocomposite-based electrochemical sensors for detecting environmental pollutants. *Ind. Eng. Chem. Res.* 60, 1112–1136. https://doi.org/10.1021/acs.iecr.0c04952

Tanaka, T., Bulinski, A., Castellon, J., Frechette, M., Gubanski, S., Kindersberger, J., Montanari, G., Nagao, M., Morshuis, P., Tanaka, Y., Pelissou, S., Vaughan, A., Ohki, Y., Reed, C., Sutton, S., Han, S., 2011. Dielectric properties of XLPE/Sio_2 nanocomposites based on CIGRE WG D1.24 cooperative test results. *IEEE Trans. Dielect. Electr. Insul.* 18, 1482–1517. https://doi.org/10.1109/TDEI.2011.6032819

Thostenson, E.T., Ren, Z., Chou, T.-W., 2001. Advances in the science and technology of carbon nanotubes and their composites: a review. *Compos. Sci. Technol.* 61, 1899–1912. https://doi.org/10.1016/S0266-3538(01)00094-X

Tong, C., 2022. Semiconducting materials for printed flexible electronics, in: *Advanced Materials for Printed Flexible Electronics, Springer Series in Materials Science*. Springer International Publishing, Cham, pp. 159–220. https://doi.org/10.1007/978-3-030-79804-8_4

Wang, L.-L., Tay, B.-K., See, K.-Y., Sun, Z., Tan, L.-K., Lua, D., 2009. Electromagnetic interference shielding effectiveness of carbon-based materials prepared by screen printing. *Carbon* 47, 1905–1910. https://doi.org/10.1016/j.carbon.2009.03.033

Wang, S.-J., Zha, J.-W., Wu, Y.-H., Ren, L., Dang, Z.-M., Wu, J., 2015. Preparation, microstructure and properties of polyethylene/alumina nanocomposites for HVDC insulation. *IEEE Trans. Dielect. Electr. Insul.* 22, 3350–3356. https://doi.org/10.1109/TDEI.2015.004903

Wang, W., Li, S., Min, D., 2016a. Enhanced flashover strength in polyethylene nanodielectrics by secondary electron emission modification. *AIP Adv.* 6, 045022. https://doi.org/10.1063/1.4948457

Wang, W., Li, S., Tang, F., Li, J., 2012. Characteristics on breakdown performance of polyethylene/silica dioxide nanocomposites, in: *2012 Annual Report Conference on Electrical Insulation and Dielectric Phenomena. Presented at the 2012 IEEE Conference on Electrical Insulation and Dielectric Phenomena - (CEIDP 2012)*, IEEE, Montreal, QC, Canada, pp. 521–524. https://doi.org/10.1109/CEIDP.2012.6378834

Wang, W., Min, D., Li, S., 2016b. Understanding the conduction and breakdown properties of polyethylene nanodielectrics: effect of deep traps. *IEEE Trans. Dielect. Electr. Insul.* 23, 564–572. https://doi.org/10.1109/TDEI.2015.004823

Wang, Y., Jing, X., 2005. Intrinsically conducting polymers for electromagnetic interference shielding. *Polym. Adv. Technol.* 16, 344–351. https://doi.org/10.1002/pat.589

Yang, D., 2012. Application of nanocomposites for supercapacitors: characteristics and properties, in: Ebrahimi, F. (Ed.), *Nanocomposites - New Trends and Developments*. InTech. https://doi.org/10.5772/50409

Yang, Y., Gupta, M.C., Dudley, K.L., Lawrence, R.W., 2005. Conductive carbon nanofiber–polymer foam structures. *Adv. Mater.* 17, 1999–2003. https://doi.org/10.1002/adma.200500615

Zazoum, B., Frechette, M., David, E., 2016. LDPE/tio_2 nanocomposites: effect of poss on structure and dielectric properties. *IEEE Trans. Dielect. Electr. Insul.* 23, 2505–2507. https://doi.org/10.1109/TDEI.2016.7736806

Zhang, L., Zhou, Y., Cui, X., Sha, Y., Le, T.H., Ye, Q., Tian, J., 2014. Effect of nanoparticle surface modification on breakdown and space charge behavior of XLPE/SiO_2 nanocomposites. *IEEE Trans. Dielect. Electr. Insul.* 21, 1554–1564. https://doi.org/10.1109/TDEI.2014.004361

Zhou, Y., Liu, F., Wang, H., 2017. Novel organic–inorganic composites with high thermal conductivity for electronic packaging applications: a key issue review. *Polym. Compos.* 38, 803–813. https://doi.org/10.1002/pc.23641

7 Functional Polymer Nanocomposites in Water Purification

Ghanshyam S. Chauhan, Sandeep Chauhan, and Kiran Kumar

7.1 INTRODUCTION: FUNCTIONAL POLYMERS AND WATER PURIFICATION

Before dealing with the functional polymer nanocomposites (FPNCs), it is desirable that functional polymers tailored for use in water purification technologies are discussed. It is necessary to clean the polluted water from municipal corporations, industrial, and other water using production processes, so that it is usable in various fields including for potable purposes. With the water pollution level reaching alarming level, different techniques are being used. There are many processes to clean the water contaminated with different pollutants such as adsorption, chemical precipitation, ion exchange, membrane separation, flocculation, or biological methods. However, the techniques mentioned in the preceding lines have issues of low practicability and operational difficulties with low efficiency, high cost, and environmental implications.

Amidst the various techniques used, adsorbents are the most desirable agents for cleaning polluted water. The adsorption process is reversible as most of the adsorbents are regenerable by removing the adsorbed pollutant entities by simple protocols. Hence, most of the adsorbents are recyclable and reusable for many cycles. Activated carbon is a commercially available effective adsorbent but with high cost and low regenerability. Among the special attributes of an adsorbent, it should have low cost, high efficiency, specificity for a pollutant, and regenerable and recyclable in multiple cycles. An adsorbent should have high sustainability. Hence, the current focus is on research to develop new adsorbents with inherent eco-friendliness. The same are inexpensive, selective, efficient, and easily regenerable. Biomass/biowaste is the most suitable feedstock material to design an adsorbent. The ease of functionalization of biomass by various modes results in adsorbents that have an inherent biodegradable nature leaving behind no waste post-use. In addition, there is a high versatility in designing the procedures for the synthesis of adsorbents. The adsorption/desorption cycles make adsorbents more cost-effective and efficient.

Functional polymers are the most important material to clean water contaminated with various pollutants including metal ions, anions, organic compounds, dyes, or generally the effluents of different industries, and those of geo-generic origin such as arsenic, nitrate, or fluoride. It must be ensured that there is no long-standing waste

 DOI: 10.1201/9781003527602-7

accumulation post-use. The synthetic polymers mainly have the issue of non-biodegradability that means waste generation post-use. Nonetheless, it is always imperative to develop effective, non-toxic, low-cost, environmentally friendly, and naturally abundant material, biowaste/biomass, for preparing new adsorbents. The biopolymer-based adsorbents have lower cost than the conventional adsorbents. Polymers, or biopolymers the main constituents of the biomass, are easy to modify to suit the tailer-made applications. The adsorption efficacy is not so high in the pristine form of synthetic or biopolymers as their functional groups as active sites are not very effective. Hence, suitable modification to incorporate more effective groups via chemical modification for better interactions with the adsorbate molecule is required (Chauhan et al., 2005, 2010, Thakur et al., 2014).

To further enhance the adsorption performance of materials used in water technology apart from the functional polymers, as discussed in the preceding part, another class of adsorbents, nanocomposites (NCs), has emerged as the widely investigated materials for use in water purification. And among the NCs, those having functional polymer matrix components have been the most effective adsorbents. These NCs are termed as polymer-functionalized nanocomposites (FPNCs).

Placed below is the structure of some commonly reported polymers used in the synthesis of FPNCs (Figure 7.1).

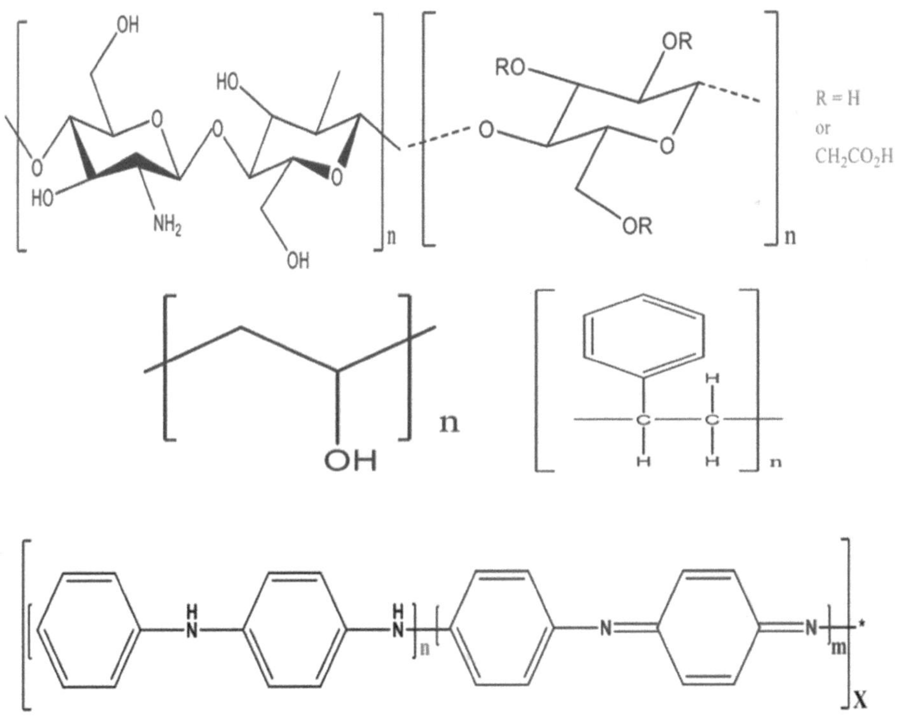

FIGURE 7.1 Examples of some functional polymers used in FPNC synthesis [from left chitosan, carboxymethyl cellulose (CMC), poly(vinyl alcohol (PVA), poly(styrene), and polyvinyl aniline (PANI)].

This chapter discusses different aspects of FPNCs from the synthesis to use in the remediation of water pollutants.

7.2 NANOCOMPOSITES

In view of their potential in the water purification field, NCs are considered as materials of the present and the future. The properties of the composite materials are unique than their components. NCs are a combination of two phases of diverse materials. Hence, these are the amassed form of two or more materials that have diverse physical and chemical properties distinguished by their interface. NCs are classified into different classes depending on the chemical composition, use of fillers, and matrix material.

NCs have a heterogenous nature with more than one component. The two phases comprising NCs are different materials. The heterogeneous nature improves the overall merit of the NCs owing to the nanomaterial used which has special physical or chemical properties. The continuous phase is termed as "matrix." The other phase in matrix is "dispersed phase." NCs has well-dispersed phase. The dispersed phase has nano-scale dimension with particle size in 10–100 nm range (Li et al., 2018).

NCs must have a well-dispersed phase that well defines the resultant properties as the dispersion and homogeneity are major issues in the outcome of the properties of NCs as the mechanical strength, chemical resistance, magnetic properties, and thermal stability along with high porosity. The heterogeneity improves overall property profile of NCs and the NPs plays an important role in that as it has special physical or chemical properties.

The structure and properties of NCs can be tailored to match the targeted applications and one such is the use of polymer as the matrix in NCs to develop FPNCs for water purification technology (Friedrich et al., 2005).

7.3 FUNCTIONAL POLYMER-BASED NANOCOMPOSITES

The NCs with functional polymers and NPs add two positive attributes, i.e. of polymer and NPs, to the final material synthesized. Hence these are referred as polymer functional nanocomposites or FPNCs. The FPNCs have varied compositions and forms. The different forms may be powder, films, or membranes. The compositions can have functional polymer such as graft copolymer, hydrogels, and NPs of different clays, metals, and inorganic oxides.

FPNC have highly complex structure is far more than the individual components considered alone. These attributes define interactions between the matrix and the NPs and result in ultimate structure. The properties of both the matrix, polymer, and the filler or reinforcement, NPs, define composite properties. The NCs often have appreciable strength, toughness, electrical, magnetic, low density, corrosion resistance, and thermal insulation (Saleh et al., 2019).

The FPNCs have a large interface resulting from the NPs and polymer matrix. The functional polymeric surface is ideal for the effective dispersion of NPs and consequent large interphase. A soft matrix results in effective reinforcement that augments FPNC properties. The resultant interface and the corresponding interphase are the determinants of the resulting properties of FPNCs. One problem that FPNCs has is that the aggregation of NPs, especially if the particles are spherical in shape. Hence, any reinforcement should improve the dispersal of NPs in matrix, resulting in better adhesion with polymer matrix (Keledi et al., 2012). Unique properties of FPNCs, both physical and chemical, are the outcome of the large surface area to volume ratio, the high interfacial reactivity of nanofillers, NPs, and appreciable mechanical properties and compatibility owing to the polymer matrix (Zhao et al., 2011). The large specific surface area of FPNCs results in high stability to the NPs. The structure of the NCs is affected by the ratio of NPs to matrix. One of the large positives of FPNCs is the stabilization of magnetic NPs. The latter, being highly reactive, have one serious limitation as these can oxidize in air. This results in the disruption of dispersion and loss of magnetic character of FPNCs. One mode of improving storage life of FPNCs is coating of NPs with polymers. That also reduces the aggregation of NPs.

The scheme for effective FPNC synthesis is provided as Figure 7.2.

FPNC can be synthesized in many forms including functional MOF. These possess multifunctional properties. These MOFs are versatile porous materials as they are high-performance materials with multifunctionalities (Unnikrishnan et al., 2021).

7.3.1 Polymers Used in FPNCs

Polymers have large molecules with large pool of active groups. Though these have significant physical and chemical properties, fillers and modifiers enhance the polymer properties for effective composite structure. The material used to improve the inherent properties of polymers include an array of nanomaterials to develop FPNCs (Zabihi et al., 2018). The polymer contributes to the FPNC synthetic process with the

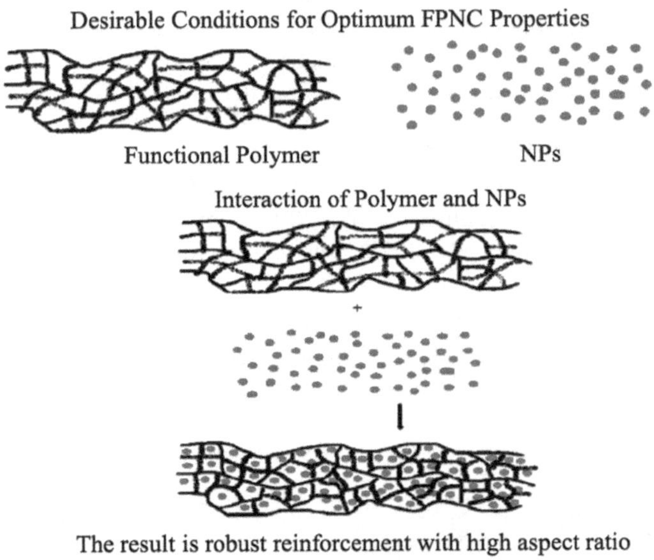

FIGURE 7.2 Schematic of NPs dispersal and reinforcement in a FPNC.

ease of processability, while the outcome of its properties is reflected in the improved stability via the NP and polymer matrix interaction. FPNCs can have positive or negative charge depending on the polymer structure. The functional polymeric surface is ideal for the effective dispersion of NPs and consequent large interphase.

To augment the properties of FPNC, it is necessary that its both components, the matrix and NPs, interact effectively (Venkateswarlu & Yoon, 2015). Effective dispersal of NPs in matrix results in excellent mechanical strength of the resultant FPNC. For this, the polymer matrix is highly a favorable support for the assembly of NCs, especially for use as adsorbents. The effective FPNCs have irreversible dispersion of NPs within the polymer. Herein, the wide dispersion of the NPs within the matrix does not aggregate much. The functional groups on the matrix add to the interactions with the adsorption process (Zhang et al., 2008).

7.3.2 Polymers-Based Classification of FPNCs

Polymer-based classification of FPNCs is depicted in the scheme presented in Figure 7.3.

7.3.2.1 Synthetic Polymer-Based FPNCs

Many synthetic polymers are used in FPNC synthesis. These include those either having the ease of functionalization by having suitable structures for functionalization or a large pool of functional groups. Hence, it has been reported that these can be further functionalized with active moieties such as imidazole (Yin et al., 2010). Some examples include polystyrene (PSty) (Sarkar et al., 2011), -:S:- based thiophene (Pty) (AL-Refai et al., 2021), -N-: based pyrrole (PPy) (Bhaumik et al., 2011a), and PANI (Ghorbani & Eisazadeh, 2013), PVA (Mbhele et al., 2003), poly(1-vinylimidazole) (Takafuji et al., 2004). These polymers have effective structural attributes that are useful to carry the dispersed NPs in the FPNC. Apart from a single polymer, the two synthetic polymers as PANI-PPy (Bhaumik et al., 2011a,b) synthetic polymers as copolymers of PP-PE and styrenebutadiene-styrene have also been reported (Chen et al., 2004).

7.3.2.2 Biopolymer-Based FPNCs

Given the sustainability drive, many biopolymer-based FPNCs have been reported. These include cellulosic (Hunne et al., 2008), chitosan (Jaiswal et al., 2012), cyclodextrin (Tarannum et al., 2022), and carboxymethylated cellulose (CMC)

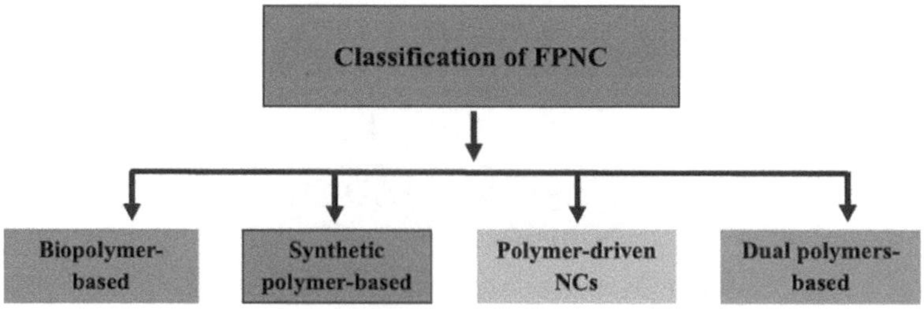

FIGURE 7.3 Polymer-based classification of FPNCs.

(Dong et al., 2011). The biopolymers possess a large pool of functional groups such as -OH, -COOH, or $-NH_2$, which offer the ease of dispersion with NPs and their easy modification to more specific and reactive species. Also, these have low costs, and having inherent biodegradability does not pose the problem of post-use waste accumulation.

7.3.2.3 Polymer-Driven NCs

Herein, polymers are used as a source to derive an active component instead of the polymer backbone as such. Polymer-derived mesoporous SiOC/ZnO NCs are a new class of material synthesized for use as adsorbents. These are prepared by a simple procedure wherein NPs get fixed onto a polymer followed by pyrolysis. One example is the synthesis of SiOC/ZnO NC by the pyrolysis of ZnO- nanopowder loaded onto vinyl-functionalized polysiloxanes and subsequent pyrolysis for 2 hat 700°C. The resultant material is amorphous in nature with a pore size <2 nm with very high specific surface area. The amount of ZnO affected surface area inversely (Hojamberdiev et al., 2012).

7.3.2.4 Dual Polymer-Based FPNCs

Apart from the single polymer composites, dual polymer composites have been reported with high stability and mechanical strength. The examples include polypyrrole (PPy) and poly(aniline) [PANI] to generate FPNC with PPy-PANI (Bhaumik et al., 2011; Javadian et al., 2014) and PPy-polyacrylonitrile (Wang et al., 2013). Such FPNCs possess more surface active groups; hence, the result is higher adsorption capacity. Similarly, two copolymers, polypropylene-ethylene and styrenebutadiene-styrene, based FPNC has been reported with $CaCO_3$ NPs (Chen et al., 2004). There are examples of two polymer composite, in the form of membrane, is the use of PAA/PVDF (Wang et al., 2008a), and PEG/nylon (Tong et al., 2011).

7.3.3 Functionalization of Polymers

The long chains and large molecular weight of polymers provide a large pool of active sites for suitable functionalization as these can be molded in different forms including films and membranes as stated in the preceding part. Polymers having active groups may be used in the native forms exemplified as poly(ethyleneimine) (PEI) or PAAc. However, to make the presence of polymers, more effective in the resultant FPNC, polymers should be functionalized with specific functional groups. For example, thiol groups were incorporated on poly(vinyl alcohol) (PVA) to generate mesoporous $(PVA)/SiO_2$ composite with thrust on the effect of PVA contents (Wu et al., 2010), PEI and tannins were used together (Huang et al., 2018). Selective active groups were generated on SiO_2 of FPNC of poly(amidoxime)/SiO_2 (Gao et al. 2010). Similarly, imprinted polymers have been reported with specific cations (Liu et al., 2011).

Polymers containing -:N, -:S: offer ease of further functionalization to create charged moieties to increase selectivity and efficacy of the reported FPNC. These include polymers having cationic and anionic functional groups ($-CH_2N^+(CH_3)_3$, etc., $-SO_3^-$) have been highly effective compared to $-CH_2Cl$ improving dispersal of nano-$Zr(HPO_4)_2$. ZrP–S showed higher capacity as Pb^{2+} ions adsorbent than the ZrP–Cl (Zhang et al., 2011).

Another attribute to widen the application spectrum of FPNCs is by enhancing its hydrophobicity. Floating FPNCs with high hydrophobicity have been reported to effectively take up oil form the polluted water (Chen et al., 2013).

7.4 DIFFERENT NP COMPONENTS IN FPNCs

The NP size and its distribution, aspect ratio, and specific surface area, affect both matrix/particle and particle/particle interactions. These form a network by interaction with the other components. The small NP content ensures their homogeneous distribution in the polymer matrix. The result is larger reinforcement. Since NP size is in the nm range, these have positive attributes for composite formation. The NPs have three-dimensional shapes for oxides and two–dimensional shapes for fibers and nanotubes. However, the shape of the NPs affects their properties especially as the spherical ones highly affect the interaction level with the other components. For uniform dispersion with low aggregation and effective interaction with the polymeric matrix, the NP: polymer ratio should be roughly 20:1. Such FPNC has strong reinforcement with high aspect ratio. Some classes of NPs used in FPNCs are discussed as follows.

7.4.1 Phosphate as NP Component

Phosphates have been used in FPNC synthesis. These affects FPNC properties resulting in fast kinetics, appreciable adsorption capacity, and high thermal stability (Vivani et al., 2008). Phosphate NPs have been combined within porous polymers to get composites. (Pan et al., 2006). $ZrPO_4$ is reported as NPs in many studies. To materialize that the NPs disperse well into the inner core of the polymeric matrix, a suspension comprising poly(styrene) and $ZrOCl_2$ was evaporated and later added into H_3PO_4 (Zhang et al., 2011). Different FPNCs were synthesized by encapsulation of nano-Zr phosphate in three microporous poly(styrene) resins having active groups viz. $-SO_3^-$, $-CH_2Cl$, and $-CH_2N^+(CH_3)_3$. Similarly, PANI-based tungstomolybdophosphate composite has been reported (Sharma et al., 2014).

7.4.2 Metal Oxides as NP Component

NPs of different metal oxides and polymers matrix-based FPNCs have been reported for water purification applications. The oxides include hydrated iron oxide NPs (Cumbal & Sengupta, 2005; Zhang et al., 2008; Sylvester et al., 2007). TiO_2 has been evaluated as a component of the NC. It has drawn a huge focus on dye adsorption and degradation (Konstantinou & Albanis, 2004). Apart from the above, Fe_3O_4, Fe_2O_3, and Fe_2O_4 have been used in FPNC. Many oxides have been used in combination with other oxides.

7.4.3 Metal Sulfides as NP Component

Sulfides have also been evaluated as a component of the NCs. An example is the use of CdS NPs for the degradation of nitroaromatics (Datta et al., 2008). The PANI-PdS-CdS-based FPNC has been reported as visible photocatalyst and anticorrosion agent (Zhang et al., 2012b). In many literature reports, CdS-based materials have been used in water splitting (Nasir et al., 2020).

7.4.4 Metals as NP Component

The FPNCs of the other NPs have been reported. These include Fe, encapsulated in polymers. The structure of the NCs is affected by the ratio of NPs to matrix. Zero-valent metal NPs used include Fe, the most used one, on (PVP) nanofibrous membranes (Xu et al., 2010) and PEG/nylon membrane (Tong et al., 2011). Fe NPs also find uses in ion exchangers (Xiong et al., 2007; Liu & Zhang, 2010). The other zero-valent metal NPs include Cu (Wu et al., 2009). These offer the ease of separation after the adsorption process. Apart from the ease of separation, without leaching out these also maintain their exceptional intrinsic physical and chemical properties. (Cumbal & SenGupta, 2005). Apart from these, various bimetallic NPs have been used. Some combinations include Fe/Pd and Zn/Pd. (Wang et al., 2008).

7.4.5 Clay as NP Component

The natural clay-based FPNCs have been reported. Apart from the easy availability with low cost, the clays have lamellar structure and high mechanical and chemical stability. The clays have lamellar structure that offers ease of FPNC formation. The clay has a negatively charged surface. It electrostatically interacts with polymer resulting in FPNC More effective the dispersion stronger the interaction between polymer and NPs. Yet, these have low surface area due to the high stacked lamellar structure that restricts accessibility of the adsorbate ions/molecules to the active sites. The FPNCs of clay overcome such limitations. Apart from the benefits of using polymers, these clays viz., bentonite, montmorillonite, and kaolinite have attracted been reported as effective adsorbents for various water contaminants. Among positive attributes for FPNC formation via effective interaction, the clay has inherent high cation exchange capacity for use as adsorbent. The clay and polymer-based FPNCs have been reported by *in situ* polymerization. The amidoxime functional groups-based monomer has been reported to form FPNC with organobentonite/polyacrylonitrile via *in situ* intercalation polymerization for use as adsorbents (Anirudhan & Ramachandran, 2008). Some examples of the polymers in FPNCs with bentonite include polyacrylamide (Zhao et al., 2010), and those with montmorillonite include modified PPy (Setshedi et al., 2013), have been reported.

7.4.6 Carbon as NP Component

The nanocarbon-conjugated FPNCs have been reported. One example includes nanocarbon colloids, and PEI has been reported for effective metal ion adsorption (Khaydarov et al., 2010).

7.5 SYNTHETIC PROTOCOLS FOR FPNCs

Apart from the essentiality of functionalization of the polymer and nature, size and % NPs used in the FPNC, the other vital characteristic in the preparation of FPNCs is the synthetic process to ensure the maximum advantages of NPs. As stated earlier also, the most of the positives of a FPNCs are ensured with high dispersal of NPs on the polymer matrix.

In FPNCs synthesis, either NPs are grafted onto the polymer backbone or polymers are anchored to NPs (Mahdavian & Mirrahimi, 2010). Nanocarbon-conjugated polymer nanocomposites (NCPC) is another technique wherein water-solubilized polymer is added with carbon colloids bind metal ions and simultaneously obtain NCPC. That results in increased NCPC size followed by precipitation procedure of extraction of the metals (Khaydarov et al., 2010). Various techniques have been reported to synthesize FPNCs. Some of these include sol–gel method, ball milling, microwave, hydrothermal process, sonication, and solution-evaporation as depicted in Figure 7.4.

Below a brief account of methods used is described (Table 7.1).

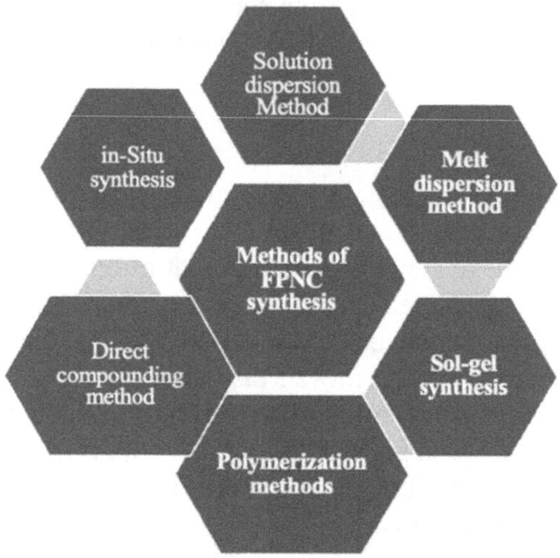

FIGURE 7.4 Graphical presentation of different synthetic methods used for FPNCs.

TABLE 7.1
Brief Details of Different Methods

S. No	Method	Remarks	References
1.	Solution dispersion	In this method, organic solvent is used to dissolve the polymer. The separately dissolved polymer and the dispersed NPs are mixed and dispersed in the polymer matrix.	Raman et al. (2012)
2.	Melt dispersion	Extrusion technique is applied to disperse and distribute nano-clay in a polymer melt phase.	Zouboulis and Katsoyiannis (2002)
3.	Sol–gel synthesis	The sol–gel method is used for the synthesis of SiO_2 matrices. It is cost-effective.	Iketani et al. (2003), Sharma et al. (2014)
4.	Microwave synthesis	Microwave synthesis is fast and economical method of FPNC synthesis.	Zabihi et al. (2020)

(Continued)

TABLE 7.1 (*Continued*)
Brief Details of Different Methods

S. No	Method	Remarks	References
5.	Polymerization methods	These methods ensure uniform distribution of NPs in the polymer matrix. The NPs get effectively dispersed and intercalated.	Mbhele et al. (2003)
6.	Direct compounding method	The direct compounding method is easy to operate and low-cost method. Polymer and NPs are compounded by solution, emulsion, fusion, or using mechanical operation.	Lofrano et al. (2016)
7.	*in situ* synthesis	Either the NPs or the polymer is generated *in situ*. To synthesize polymers, *in situ* monomer/s and the NPs are dispersed along with initiator and solvent. Or NPs precursor is used which generates NPs within the polymer matrix.	Wang et al. (2011)
8.	Template method	The template synthesis uses nanoporous materials to make NPs of different shapes, so that nm-scale tubes or fibers can be synthesized from a variety of materials including conducting polymers.	Shukla et al. (2010)

7.5.1 Solution Dispersion Method

Herein, organic solvent is chosen which can dissolve the polymer and well disperse the NPs. The separately dissolved polymer and the dispersed NPs are thoroughly mixed to ensure dispersion of NPs in polymer matrix. The removal of solvent is the limitation of this protocol. The process may involve some methods those are combined at different phases of NCs the formation. The methods include hydrothermal co-precipitation and solvo-thermal. These may be combined and the metal can be reduced using a reducing agent (Raman et al., 2012).

7.5.2 Melt Dispersion Method

Polymers, generally the thermoplastics, when heated flow, and on cooling these harden. The melt phase of the polymer is used for the effective dispersal of NPs, when the polymer melt is subjected to molding. The extrusion method is used for the same. The thermoplastics are easily remold and recycled to new products. One example of FPNC synthesized via this protocol is that of PE powder and CNTs (Mahfuz, et al. 2005).

7.5.3 Sol–Gel Method

This method is used to synthesize SiO_2-based matrix. This method is apposite to synthesize hybrid composites using low energy. The silicon alkoxides are used to generate a sol–gel matrix as exemplified by the SiO_2/C. The method is low cost (Iketani et al., 2003). The PANI-based tungsten molybdenum phosphate composite was synthesized via sol–gel technique (Sharma et al., 2014).

7.5.4 Microwave Synthesis

Microwave synthesis technique is employed in the synthesis of NCs. The method is fast and economical, but also product specific. These aspects make it a potential method to synthesize FPNCs (Zabihi et al., 2020).

7.5.5 Polymerization Methods

The structure, especially the surface of NPs plays significant role that results in an effective FPNC. Similarly, the binding capacity of polymer surface, having effective active sites, is also a vital factor (Lofrano et al., 2016). These two aspects of FPNC synthesis are highly materialized by choosing appropriate polymerization methods for the synthesis of PNFCs. The synthetic protocol should result in a uniform and effective dispersal of NPs in polymer matrix. The NPs get effectively dispersed and intercalated. Hence, the resulting properties are regulated. The interphase interaction between NPs and polymer when effective generate a FPNC that is of high applied interest (Mbhele et al., 2003). The nature of polymer used, synthetic polymer or a biopolymer, differentiate these as two different categories as discussed below.

7.5.6 Direct Compounding

The direct compounding method is easy to operate and low cost method. It has been widely used to synthesize FPNCs (Lofrano et al., 2016). Polymer and NPs are compounded by various methods including emulsion, solution, or mechanical operation. However, direct compounding degrades the polymer via melt compounding. It also results aggregation of NPs to large lumps. The result is lowering the advantages of their nano-scale dimension. To avoid the same, NPs are surface modified. That results to retain their effective dispersal in polymer matrices. The cellulose/TiO_2 FPNCs have been synthesized. Cellulose has highly organized structure with its chains joined tightly by hydrogen bonding. The later improves contacts between the cellulose chains and the NPs.

7.5.7 In Situ Synthesis

In *in situ* approach, either the NPs or the polymer is generated *in situ*. To synthesize polymers *in situ*, monomer/s and the NPs are dispersed along with initiator and solvent. The same gets polymerized to generate the FPNC (Wang et al., 2011). For synthesis approach, NPs precursor is used which generates NPs within the polymer matrix (Wu et al., 2006). This approach is that the precursors of NPs and the monomers are mixed in solvent with an initiator. The blending results in the simultaneous preparation of NP_S and polymer, resulting in FPNC formation.

7.5.8 Template Method

The template synthesis involves use of material with nanopores. It is used as template to make NPs of different shapes. The latter is a significant feature of the template method as nm-scale tubes or fibers can be synthesized from a variety of materials including conducting polymer and carbons. The use of porous alumina template has been reported to synthesize nanoarray FPNC of gold and polymer (Shukla et al., 2010).

7.6 IMPORTANT STRUCTURAL FEATURES OF FPNCs AS ADSORBENTS

FPNCs meet many essential attributes of high value adsorbents with high sustainability. These are attractive materials as adsorbents having many structural aspects those are propelling new researches in this area. Moreover, these are low cost. Hence, FPNCs have been mainly fabricated for use as adsorbents from wastewater (Cumbal & Sengupta 2005; DeMarco et al., 2003; Ponder et al., 2000; Sylvester et al., 2007).

These have become highly promising materials with low cost. The sustainable sources as agricultural residues, inorganic materials, and biopolymers for adsorption of ions/dyes.

Furthermore, magnetic characteristic of FPNCs adds to the ease of separation from the adsorbate solution by simple protocols. Owing to the different structural, cost, ease of operation, and environmental upsides FPNCs have become most sought after materials for water purification processes. The polymer backbone can be tuned for specific applications with either exchangeable cation attached to the anionic moieties, such as sulphonic groups, or cationic moieties such as N^+ with counter anions. The former acts cation exchangers while the later acts as anion exchangers.

7.6.1 Structural Attributes of a FPNC for Use as an Adsorbent

FPNCs have large surface area with high pore volume. Along with that polymers have large pool of effective binding sites. The augmented surface area and effective surface morphology results in a higher sorption capacity for contaminant removal. The high efficacy toward the adsorbate can be explained based on their structural attributes. The nature of polymer, synthetic protocol, and the extent and uniformity of NPs dispersion on polymer surface are high attributes that makes these useful materials in the adsorption of different adsorbates. The charge on the surface of the FPNCs is also important which is provided by charged functional groups, present on the polymer surface, such as $-N^+$, $-COO^-$, and $-SO_3^-$. Such charge on the surface of polymer also contribute to the high NPs dispersion with the FPNCs to enhance the adsorption efficiency (Figure 7.5). FPNCs have positively or negatively charge that facilitate their use as ion exchanger (Ghorbani & Eisazadeh, 2013).

7.6.2 Strategizing Structural Attributes of FPNCs via Synthetic Protocols

As stated in the preceding part, FPNCs have large surface area and porous nature. These attributes are important along with the large pool of adsorption sites (Saleh et al., 2019). Many of these attributes arise from the polymer used as matrix. The polymer matrix also contributes positively with the ease of high processability and flexibility to the FPNCs. The NPs have a high aspect ratio and that along with large surface to volume ratio. These aspects strengthen the polymeric matrix. Hence, the suitable and

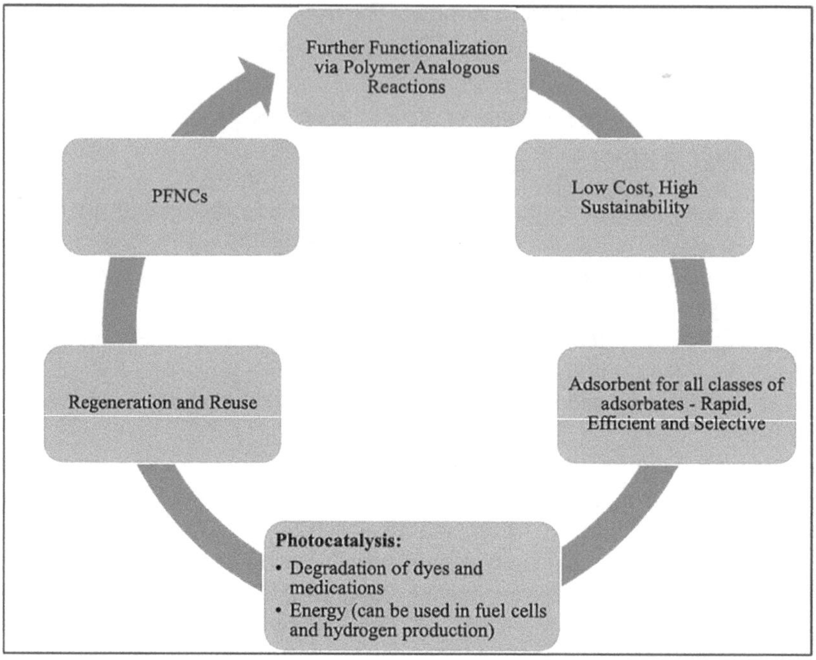

FIGURE 7.5 Important structural requisites and properties of FPNCs as adsorbents.

active surface of FPNCs for defined applications, with mechanical strength, have been perceived as the positive attributes of a FPNC for use as an adsorbent (Ghorbani & Eisazadeh, 2013). It is of high importance in the FPNC synthesis that there exists high possibilities that different synthetic modes can be used to manipulate the targeted properties of FPNC by choosing an apposite synthetic mode. One such aspect is ratio or %magnetic core (NPs) to polymer which plays a crucial aspect in FPNC synthesis. Hence, to tailor the apposite properties of an FPNC, the NPs to polymer ratio is a crucial factor. Especially the amount of the polymer highly affects FPNC properties. The lower amount of the polymer decreases the adsorption capacity as active sites for NPs dispersion and adsorption are reduced. However, such relationship is also adsorbate specific. While the Hg^{2+} ions adsorption capacity increased with the Fe_3O_4 amount to a certain extent, but decreased on high increase of the same. The polymer having the specific groups showers specificity on the adsorption property as thiol groups are highly specific for the Hg^{2+} ions. The Fe^{2+} ions are specific for the arsenic removal as the iron contents affect the arsenic adsorption capacity.

7.6.3 Further Functionalization of FPNCs via Polymer Analogous Reactions

The polymer offers the ease of further functionalization via simple polymer reactions. For example, the $-NH_2$ groups of a polymer act as cation adsorbent, but these can be further functionalized with ease to generate $-NH_3^+$ groups that function as anion exchanger post-quaternization interaction. The charged surfaces are more effective than the neutral precursors. The :NH_2 groups on the amino-functionalized adsorbents are effective for cation adsorption as these can coordinate with the metal ions via lone pair of electrons.

7.6.4 Imparting Selectivity on FPNC

The choice of a polymer or functional groups on it can impart selectivity on the polymer; hence the FPNC. Functional groups such as thiol or amidoxime are very effective and selective for the specific adsorbate. The use of Fe^{2+} NPs also is effective and selective for As(V) anions. The charged functional groups covalently bound on the polymeric matrix attract the oppositely charged contaminants, for permeation, present in the polluted water. This phenomenon has been linked to the Donnan membrane principle (Zhang et al., 2008).

7.6.5 Cost Evaluation of FPNC as Adsorbent

The cost of an adsorbent is the key factor to evaluate its sustainability. Hence, cost analysis of an adsorbent is important. The preceding discussion relates to all desirable aspects of FPNC as adsorbent. These factors need to be assessed while evaluating the cost of the adsorbent adding the cost of the components and the synthetic process used to the performance of the adsorbent. The cost cannot be assessed mainly from the FPNC components such as polymer, crosslinker, or NPs. Instead, it must be linked to the rapidity, selectivity, and adsorption efficiency of the FPNC, thereby a more efficient adsorbent with a high regeneration and reusability factor will prove to be low cost despite the high cost of the FPNC components (Zhou et al., 2009b).

7.7 FPNCs AS METAL IONS ADSORBENTS

This section deals with some specific FPNCs for use in metal ion adsorption. The FPNCs have been widely reported as metal ions adsorbents (Meric, 2016). The focus has been laid on some new classes of FPNCs especially those having -N: and -:S: based adsorbents.

7.7.1 PANI and Other: N- and :S-Based Polymers as Adsorbents

These FPNCs can be easily synthesized. Furthermore, these can be functionalized with ease to make these more effective and selective for the defined adsorption applications. To fabricate adsorbents for use as adsorbents, poly(aniline) (PANI) is a highly attractive polymer to get FPNCs. It offers the ease of further modification to improve efficacy and selectivity in the adsorption process. Apart from PANI, polythiophene (PTh), PPy and PEI are attractive candidates for this purpose. The presence of -N or/ and -S atoms on the functional groups present on the polymers are very effective in adsorption of the adsorbate moieties. However, in this study, it was reported that the structure of polymer affects adsorption capacity as PPy (-:N containing) has considerably higher adsorption capacity than the S containing PTh (Zare et al., 2018).

7.7.2 Graphene Oxide-Based Adsorbents

The graphene oxide/Fe_3O_4 FPNCs modified with poly(acrylic acid) (PAAc), PAAc/GO/Fe_3O_4, possess magnetic property with water solubility offering the ease of separation for reuse and capability of interaction with the adsorbate ions. It showed high adsorption efficiency for Cu^{2+}, Cd^{2+}, and Pb^{2+} ions from their aqueous solutions with appreciable removal efficacy of ~85% after five cycles (Zhang et al., 2013). The graphene-TiO_2 has been reported to be effective in reducing Cr^{6+} ions to Cr^{3+} ions on exposure to sunlight (Zhang et al., 2012a). TiO_2 NPs were uniformly anchored on graphene sheets via esterification reaction.

7.7.3 Metal Ions Adsorbents with Nanocarbon

The nanocarbon colloids (NCC) and PEI-based FPNC were evaluated as adsorbents for different ions viz., Zn^{2+}, Cd^{2+}, Cu^{2+}, Hg^{2+}, Ni^{2+}, and $Cr^{6}+$ ions. The removal/adsorption efficiency was higher than >99%. The determinants of the adsorption have been reported in both the structural aspects of the FPNC and adsorption parameters. The former includes the NCPC synthesis process, the size of NCC, and the ratio of NCC to PEI in FPNC (Khaydarov et al., 2010).

7.7.4 CNTs-Based Metal Ion adsorbent

CNT-based FPNCs have been used as adsorbents using various inorganic oxides for arsenic, Cr^{3+}, Zn^{2+}, Cd^{2+}, and Pb^{2+} (Saleh, 2015a). In another study, silica-C nanotubes were used as adsorbents and compared with silica/AC (Saleh, 2015b). However, the carbon-based adsorbents are not very efficient as adsorbents for heavy metal ions than the other ones.

7.7.5 Zero-Valent Metal NPs-Based FPNCs

There are numerous reports on the use of zero-valent metal NPs in PNFCs. Amidst these, Fe0 has been commonly reported to develop effective adsorbent for different applications (Ponder et al., 2000; Xiong et al., 2007). Similarly, Cu0 (Jaiswal et al., 2012) and Zn0 (Song et al., 2008) have been reported as effective part of the FPNC for different applications.

7.7.6 Biopolymers-Based FPNCs for Metal Ions

Various biopolymers have been reported as FPNC components for FPNCs and evaluated as adsorbents of the heavy metal ions. Some examples include those based on chitosan, carboxymethylcellulose, and cyclodextrin. The chitosan-based magnetic microspheres modified with thiourea showed high capacity for the adsorption Cu^{2+}, Hg^{2+}, and Ni^{2+} ions (Zhou et al., 2009a). Compared to Cu^{2+} and Ni^{2+} ions, the adsorption capacity was too high for Hg^{2+} ions. Such trends can be ascribed to the presence of thiourea as -S- has far higher affinity for Hg^{2+} ions compared to the other ones. Cyclodextrin-based magnetic composites, wherein β-cyclodextrin was

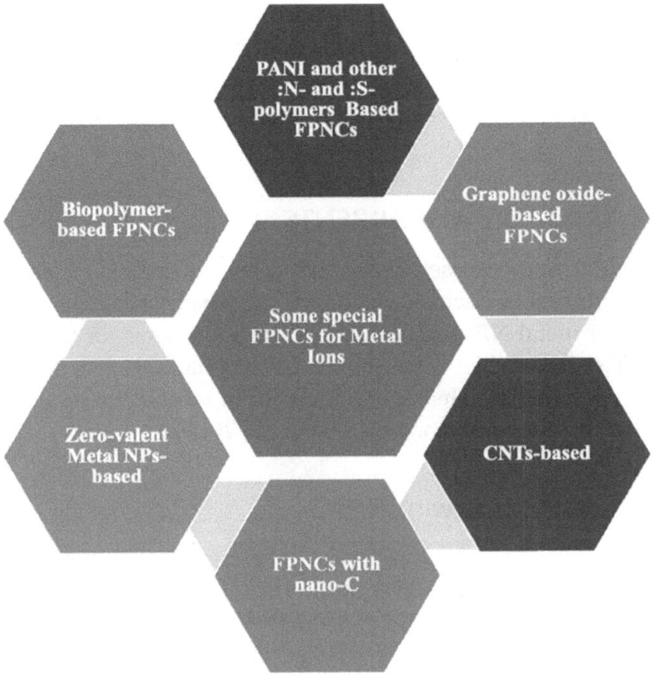

FIGURE 7.6 Some of the functional polymers with high positives for use as adsorbents.

functionalized via carboxymethylation, and Fe_3O_4 NPs were used on such synthesized matrix (Figure 7.6). It also showed selective adsorption of Pb^{2+}, Cd^{2+}, and Ni^{2+} ions in competitive adsorption experiments. The results were the highest for Pb^{2+}ions (Badruddoza et al., 2013).

7.7.7 Two or Three Polymers-Based FPNCs

The presence of two polymers has been reported to be more effective than the single polymer in FPNCs. The adsorption efficacy arises from the presence of more abundant surface groups. Apart from that higher stability is attained than the single polymer FPNC (Wu et al., 2016). PPy is a apposite polymer that combines with the polymer counterpart as PANI (Javadian et al., 2014) or polyacrylonitrile (Wang et al., 2013). The PANI/PPy was found appreciably efficient and rapid Cd^{2+} ions adsorbent. Likewise, PPy-PANI nanofibers showed high adsorption capacity for Cr^{6+} ions. Moreover, it was reported to act as catalytic agent to reduce Cr^{6+} to Cr^{3+} ions (Bhaumik et al., 2012). The PANI/PPy-based FPNC was coated on rice husk ash, and the composite was evaluated for the removal of various pollutants from the textile wastewater (Ghorbani & Eisazadeh, 2013). The PPy/polyacrylonitrile nanofibers with core–shell structure have been reported. These showed rapid and efficient adsorption of Cr^{6+}, along with recyclability (Wang et al., 2013). Lignin, the hugely available biopolymer was utilized to fabricate a hybrid FPNC with a copolymer of poly(3,4-ethylenedioxythiophene)/polystyrene sulfonate (PEDOT/PSS). It showed high adsorption capacity for Pb^{2+} ions. This example illustrates the relevance of adding another polymer to generate an effective adsorbent as the already appreciably high adsorption capacity of PEDOT/PSS was nearly increased two-fold in case of PEDOT/PSS/Lignin (Checkol et al., 2018).

7.8 FPNCs AS ANION ADSORBENTS

As stated earlier also, the capacity of an adsorbent for cations or anion removal is subject to the nature of surface charge on the adsorbent or nature of active groups. The metal ions or metal NPs as such also act as active sites for the anions (Kumar et al., 2009). The immobilized Fe0-based FPNCs have been widely reported for environmental applications (Huang et al., 2008). In most of these adsorbents, Fe or FeO-based FPNCs have been reported (Pan et al., 2009).

7.8.1 Phosphate Removal with FPNCs

Anion exchanger FPNC has been reported to be effective in phosphate removal. A hybrid anion exchanger was effective at the trace level removal of phosphate from the wastewater (Blaney et al., 2007). A hydrate FeO has also been used in an FPNC to enhance the removal of phosphate from polluted water (Pan et al., 2009). Using commercial FPNC, HFO-201, phosphate has also been effectively removed from real samples from municipal wastewater (Hua et al., 2013).

7.8.2 Arsenic Removal with FPNCs

There are many reports wherein FPNCs including PANI-based were evaluated to remove as ions from water. This is based on geogenic nature of the Fe-As. These NPs of or as its oxide have been mostly used as such (Katsoyiannis & Zouboulis, 2002) or in the hydrated form (Cumbal & Sengupta, 2005; DeMarco et al., 2003; Sylvester et al., 2007). These studies have been also related to the role of Donnan membrane effect (Cumbal & Sengupta, 2005; Zhang et al., 2008). Biopolymers have also been reported as matrix for FPNCs synthesis with targeted adsorption of as ions as reported for the Fe-loaded alginate beads as adsorbent (Zouboulis & Katsoyiannis, 2002).

7.8.3 Fluoride Removal with FPNCs

The use of FPNCs for F^- removal is exemplified by the PPy/Fe_3O_4 adsorbent (Bhaumik et al., 2011a). PEI was linked to the hydrous ZrO and later immobilized on the PAN nanofibers. The resulting FPNC was used as an effective adsorbent for F^- removal at a wide range of pH (Wang et al., 2019). An HZO-modified PPy-based adsorbent had high fluoride affinity with appreciable regeneration and reusability (Li et al., 2024). Graphene-based FPNC has been reported with high adsorption capacity (Mohan et al., 2017).

7.9 FPNCs AS ADSORBENT-CUM-CATALYSTS TO DEGRADE VARIOUS POLLUTANTS

The adsorbent attributes of the FPNCs do not only remove effluents, cations, and anions from the contaminated water via the adsorption process but also include many emerging in polluted water. However, these also can be tuned to act as catalysts to degrade the targeted contaminants as the adsorption is followed by the catalytic degradation process. These attributes are linked to not only the catalytic efficiency but also the adsorption efficiency, which is the primary factor in the catalysis process. For the high photocatalytic activity of NPs, such as TiO_2, anchored on the polymer, the latter should have antioxidative properties on exposure to UV/visible light radiation. Mostly, polymers used have saturated C-chain. The other class includes fluoropolymers, or poly(dimethylsiloxane) (PDMS). The more the adsorption efficiency of the adsorbent, the more the ease of effective degradation of the adsorbed entities. The NPs used in the degradation of various pollutants include different oxides such as TiO_2 (Konstantinou & Albanis, 2004) or as a combination of the two oxides as WO_3/TiO_2 (Kwon et al., 2000). The WO_3-loaded TiO_2 was studied to assess its photocatalytic behavior in oxidation reaction. Photocatalytic activity of FPNCs can be attained in various forms including films as reported for the TiO_2/PDMS hybrid films (Iketani et al., 2003). Sulfides such as CdS synthesized by the temperature tunability of size showed that it has efficacy in the photocatalytic degradation of nitroaromatics was size dependent (Datta et al., 2008).

Degradation of some classes of the pollutants, including the emerging ones, are discussed below those have been reported to be adsorbed, and degraded by the FPNCs.

7.9.1 Degradation of Dyes

Dyes are major contaminants of industrial water. Hence, their removal via adsorption to enable photocatalytic degradation is necessary. Various oxides based NCs have been reported to degrade azo dyes (Daneshvar et al., 2004; Konstantinou & Albanis, 2004). FPNCs based on chitosan and nano-CdS have been reported as photocatalysts for decolorization, and degradation of Congo red in the presence of visible light (Zhu et al., 2009). The degradation takes place via breaking the N=N bonds and the aromatic part. The dye was more decolorized under the acidic than the alkaline conditions. The NO_3^- ions augmented degradation, but other anions studied had an inhibitory effect with that of Cl^- being the most marked. The FPNC had good stability and showed regeneration and recyclability. Acid-supported resin with Fe0 has been effective in acid blue 113 azo dye (Shu et al., 2010). For polymer-derived mesoporous SiOC/ZnO nanocomposite the MB, cationic dye, adsorption decreased with ZnO increase. MB was degraded by the presence of ZnO nanopowders on the matrix (Hojamberdiev et al., 2012). Methyl orange, anionic dye, was degraded with poly(3-hexylthiophene)-modified TiO_2 (Wang et al., 2009). Dye degradation has also been studied as a function of different NPs viz. SiO_2 or TiO_2 on the same matrix poly(1-naphthylamine) (PNA). The pristine PNA showed just

9% degradation. The dye degradation was appreciable with TiO_2 ~60% and low of ~28% with SiO_2 (Ameen et al., 2011).

7.9.2 Degradation of Organic Pollutants

WO_3 has been reported as the catalyst to degrade cyclopentene to glutaraldehyde (Lu et al., 2011). As stated earlier also, sulfides such as CdS have been reported to photocatalytically degrade nitroaromatics (Datta et al., 2008). The other nitroaromatics, para-nitrochlorobenzene, have also been reported to degrade via catalytic reduction with the dual Fe–Pd NPs loaded onto sodium CMC (Dong et al., 2011). The FPNC was highly affective and rapid by degrading 100% of 50 ppm within 40 min. The role of bimetallic particles in the FPNC has been reported for the Pd/Sn supported on resin. The FPNC has been reported to effectively dechlorinate carbon tetrachloride (Lin et al., 2005), and trichloroethylene in aqueous medium via the reductive pathway (Lin et al., 2009).

The FPNC based on the zero-valent Fe contained in the porous carbons driven from the waste biomass acted as efficient adsorbent and dechlorination catalysts for the PCBs (Liu & Zhang, 2010); Varanasi et al., 2007).

The zero-valent Zn-based NC has been reported to reduce and dechlorinate tetrachloroethylene (Song et al., 2008). Similarly, 1,2,3,4-TCDD, a tricyclic aromatic with four -Cl groups, has been dechlorinated by using the Zn0-based FPNCs (Wang et al., 2008b).

7.9.3 Degradation of Herbicides and Pesticides

Herbicides and pesticides are serious contaminants in water bodies, and these creep in the food chain. These are organic compounds with chlorine or phosphorous components. These include organochlorine or organophosphorous insecticides. The TiO_2 suspensions have been reported to be effective in photocatalytically degrading s-triazine organophosphorus insecticides and herbicides (Konstantinou et al., 2001). Similarly, there are many reports wherein FPNCs have been reported for this purpose. FPNCs based on zero-valent copper-chitosan FPNC for the removal of organophosphorous pesticides, malathion, parathion, and methyl parathion are present in the agricultural overflow (Jaiswal et al., 2012).

7.9.4 Degradation of Other Pollutants

Chlorinated organic compounds are one of the most common contaminants in polluted water. One of these is perchlorate Cl(VII), which is a challenging pollutant in water sources. It can be degraded to chloride with FPNCs having zero-valent Fe NPs. The outcome has been reported rapidly and with very high efficiency. These FNPCs have been reported with stabilized Fe0 NPs with starch and CMC. The high efficiency in degradation and rapid degradation rate are the two important aspects of this study as perchlorate was completely degraded in samples prepared from the contaminated water as well as ion-exchange brine (Xiong et al., 2007).

Similarly, Fe0 NPs are highly efficient in Cr(VI) reduction when used with the CMC-stabilized Fe0 NPs. The latter showed low degree of agglomeration, but Cr(VI) reduction was very high compared to the ones without using a stabilizer (Wu et al., 2009). The polystyrene-supported Fe0 NPs has been very effective agents for nitrate reduction. This study elaborated and focused on the role of functional groups on the surface of the polymer (Jinag, et al., 2011). The degradation studies of some organic pollutants are presented in Table 7.2.

TABLE 7.2
Degradation of Various Water Pollutants with FPNCs

S. No	FPNC	Pollutant	Special Feature	References
	Degradation of Chlorinated Organic Compounds			
1.	Fe0-immobilized on cationic exchange membrane	Trichloroethylene	FPNC reusable repeatedly after re-reduction.	Kim et al. (2008)
2.	Supported Pd/Sn bimetallic nanoparticles	Trichloroethylene	FPNC performed ~2 times faster than the commercial one.	Lin et al. (2009)
3.	PAA/PVDF membrane-immobilized Pd/Fe nanoparticles	Trichloroacetic acid		Wang et al. (2008)
4.	Cu0-cationic resin	Carbon tetrachloride	Reducing the size of the copper particles hugely increased reactivity.	Lin et al. (2005)
	Degradation of Nitroaromatics			
5.	Fe0-loaded PEG/nylon membrane	Nitrobenzene	Rapid rate of contaminant degradation.	Tong et al. (2011)
6.	Fe–Pd NPs loaded onto sodium CMC	para-Nitrochlorobenzene	Effective and rapid degradation.	Dong et al. (2011)
7.	Degradation of Dyes			
8.	Chitosan-nano-CdS	Congo red dye	N=N bonds breaking and aromatic fragment degradation with reusability.	Zhu et al. (2009)
9.	Resin supported nano-Fe0	Acid blue 113 azo dye	Using resin supported nano zero-valent iron	Shu et al. (2010)
10.	SiOC/ZnO	Methylene blue	ZnO nanopowder significantly enhanced MB degradation.	Hojamberdiev et al. (2012)
11.	Poly(3-hexylthiophene)-modified TiO_2	Methyl orange	Oxidative degradation under visible light, high stability after reusability for 10 cycles.	Wang et al. (2009)

(Continued)

TABLE 7.2 (*Continued*)
Degradation of Various Water Pollutants with FPNCs

S. No	FPNC	Pollutant	Special Feature	References
12.	SiO_2 or TiO_2 on the poly (1-naphthylamine) (PNA)	Dye	TiO_2 far effective than SiO_2	Ameen et al. (2011)
		Degradation of Herbicides and Pesticides		
13.	Cu0-chitosan	Organophosphorous pesticides, malathion, parathion, and methyl parathion	High adsorption capacity with appreciable degradability of pesticides.	Jaiswal et al. (2012)
	Degradation of Inorganic Ions			
15.	CMC-stabilized Fe0	Cr(VI) reduction	The Cr(VI) reduction very high than the ones without using a stabilizer.	Wu et al. (2009)
16.	PSty-Fe0	Nitrate reduction	The role of functional groups on the surface of the polymer signified.	Jiang et al. (2011)
17.	Fe0 with starch and CMC	Percholorate ions	Rapid and complete degradation.	Xiong et al. (2007)

7.10 FUNCTIONAL POLYMERS-NANOCOMPOSITES-BASED MEMBRANES

Membrane fouling is the main drawback of the membrane technology in water purification. To overcome the problem of fouling, the loading of NPs onto membranes has been studied. As aforesaid, NPs impart high strength to the polymeric membrane owing to the high aspect ratio, large pool of functional groups or active sites, and surface to volume ratio. These have attributes such as large surface area owing to the loaded NPs. Apart from the adsorption behavior these catalyse the adsorbate degradation process as also discussed in the previous section for various FPNCs. The membranes with NPs loaded have been termed as the "advanced composite membranes". NPs imparts permeability control and fouling resistance.

7.10.1 Polymers Used in FPNC Membranes

The polymer used in membranes include polyvinylidene fluoride, polysulfone, and polyether sulfone. All these are highly appropriate polymers for water treatment. The effect of hydrophobicity has been studied for anti-fouling properties (Saeki et al., 2016). Among others, most reported polymers in these membranes are PSty (Alghunaimi et al., 2019), PE (Mahfuz et al., 2005), PCL (Jimenez et al., 1997), PP (Kawasumi et al., 1997), PVA (Mbhele et al., 2003), and PVP (Xu et al., 2010). These have improved performance over conventional membranes along with the high thermal and chemical resistance. The effect of hydrophobicity, resulting from the hydrophobic nature of the polymers used, is a significant attribute that affects fouling. The photocatalytically active membranes having PVA and TiO_2 have been synthesized via electrospinning (Costa et al., 2010). The NPs have been functionalized with active functional groups. Such NPs enhance the desirable attributes of the FPNCs such as high selectivity, affinity to matrix and adsorbate, and reactivity of the NPs with the contaminants in the polluted water. These aspects significantly advance the output of the NPs in various applications (Jankeviciute et al., 2013).

7.10.2 NPs Loaded on Membranes

The NPs used on membranes include Ag, zeolites, TiO_2, SiO_2, Ag, and CNT (Mehwish et al., 2014) or bimetallic NPs (Saleh, 2018). Other NPs with high surface porosity have also been used in the membranes. Nanozeolites, posing both porosity and hydrophilicity, reported in many studies and salt rejection of membranes. Among others, Fe, CuO, CdS, ZnS, FeO, ZnO, and SiO_2 have been reported (Jin & Maduraiveeran, 2017).

Graphene nanomaterials along with high surface area have many positive attributes required to develop functional materials in membranes. These include substantial strength, thermal stability, and conductivity. The membrane has been synthesized wherein graphene was grafted with long chain acid, 9-octadecenoic acid, and modified with PSty for effective removal of oil from water (Alghunaimi et al., 2019). CNT is also loaded on different polymers. It can also impart in anti-fouling properties to the membrane.

7.10.3 Some Applications of the FPNC Membranes

As stated earlier, FPNC or advanced membranes have improved permeability and salt rejection (Qu et al., 2013). These also have low fouling. Some applications of FPNC membranes are discussed below.

7.10.3.1 Reduction of Organic Pollutants

Reduction of nitrobenzene has been reported with Fe0-loaded PEG/nylon membrane. (Tong et al., 2011). Pd/Fe-loaded-PAAc/PVDF membranes have been shown to dechlorinate CCl_3COOH (Wang et al., 2008). The FPNC membrane have been effective in degradation of halogenated aliphatic compounds (Xu et al., 2010). Fe0 encapsulated in cation exchange membrane have also been reported effective in the degradation of trichloroethylene (Kim et al., 2008).

7.10.3.2 Anti-fouling Properties

Apart from the adsorption and degradation positive aspects of the FPNC membranes, microbicidal properties are of high significance to clean the microbes-infected water and inhibit membrane fouling. For example, CNTs loaded on different polymers impart in anti-fouling and self-cleaning properties to the membrane (Zhu et al., 2006). TiO_2 NPs have been reported to have appreciable antibacterial activity. Loading these on PP-based membranes has been reported to show high antibacterial activity, and it can be used to clean water of different bacteria (Kubacka et al., 2009).

7.11 SEPARATION, REGENERATION, AND REUSE OF FPNCs

One of the effective attributes of any adsorbent is its' ease of regeneration following green protocols. Furthermore, it is also highly necessary to reuse the regenerated adsorbent time and again, in multiple cycles. The ease of regeneration and multiple recyclability are essential to reduce the cost of the adsorbent used. The FPNCs have regeneration and reusable nature (Zhou et al., 2009). However, the ease of separation of the adsorbent from the solution is very important. In this section, separation, regeneration, and reusability have been discussed.

7.11.1 Separation of FPNC Adsorbents

In view of the economic and environmental concerns, the adsorbents must be separable post-operation. The separation processes employed should be low energy consuming and economical. Generally, centrifugation or filtration is employed for the separation of the adsorbent. The process of using a basic or acid solution has its shortcomings. However, given the magnetic nature of most of the NPs on the FPNCs, the adsorbate-loaded adsorbents are separable by simple settlement. Hence, it is most suitable in view of the energy and economic requirements of the separation process. Moreover, this process helps in retaining much of the structure integrity of the adsorbent which otherwise get reduced or totally lost under chemical environment such as pH of the solution (Yavuz et al., 2006).

7.11.2 Regeneration and Reusability

In many reported cases, the retention of high adsorption capacity has been in multiple cycles. An example is the Cr(VI) reduction and ion exchange, wherein ~80% adsorption capacity was retained after five repeat cycles (Wang et al., 2013).

7.12 CONCLUSIONS AND FUTURE PROSPECTIVES

This is a comprehensive account of FPNC synthesis, adsorption attributes, and performance against different adsorbates. Most relevant studies were used to develop the text of different sections. Compared to the non-composite adsorbents, these are very attractive materials for use in real water situation. These materials have emerged as the future with wide spectrum applications against different pollutants including metal ions, anions, dyes, herbicides, and pesticides with high positive of pollutant degradation via the photocatalytic path. These are rapid and selective for the applications applied. Ease of regeneration and reusability are other aspects that add value to these adsorbents. However, the regeneration is not viable in multiple cycles. This aspect needs to be improved in these materials. Many materials have been reported having biocidal properties that is requisite for treating potable water. However, the components of the adsorbents should be safe as the danger of exposure needs to be addressed. Despite the necessity for removing the preceding limitations, polymer nanocomposites are the materials of the present and the future. More importantly, green and sustainable aspects need to be addressed with a high focus on lowering the cost of production, enhancing the ease of synthesis, having high efficacy in the application applied for, and adhering to zero-waste generation post-use. These aspects can be considered for using biopolymers or biomass waste as the matrix.

REFERENCES

Alghunaimi, F.I., Alsaeed, D.J., Harith, A.M., Saleh, T.A. (2019). Synthesis of 9- octadecenoic acid grafted graphene modified with polystyrene for efficient light oil removal from water. *Journal of Cleaner Production*, *233*, 946–953.

AL-Refai, H.H., Ganash, A.A., Hussein, M.A. (2021). Polythiophene and its derivatives–based nanocomposites in electrochemical sensing: A mini review. *Materials Today Communications*, *26*, 101935.

Ameen, S., Akhtar, M.S., Kim, Y.S., Shin, H.S. (2011). Nanocomposites of poly(1-naphthylamine)/SiO_2 and poly(1-naphthylamine)/TiO_2: Comparative photocatalytic activity evaluation towards methylene blue dye. *Applied Catalysis B*: Environmental, *103*(1–2), 136–142.

Anirudhan, T., Ramachandran, M. (2008). Synthesis and characterization of amidoximated polyacrylonitrile/organobentonite composite for Cu(II), Zn(II), and Cd(II) adsorption from aqueous solutions and industry wastewaters. *Industrial Engineering Chemistry Research*, *47*, 6175–6184.

Badruddoza, A.Z.M., Shawon, Z.B.Z., Tay, W.J.D., Hidajat, K., Uddin, M.S. (2013). Fe_3O_4/cyclodextrin polymer nanocomposites for selective heavy metals removal from industrial wastewater. *Carbohydrate Polymers*, *91*, 322–332.

Blaney, L.M., Cinar, S., SenGupta, A.K. (2007). Hybrid anion exchanger for trace phosphate removal from water and wastewater. *Water Research*, *41* 1603–1613.

Bhaumik, M., Leswifi, T. Y., Maity, A., Srinivasu, V., Onyango, M.S. (2011a). Removal of fluoride from aqueous solution by polypyrrole/Fe_3O_4 magnetic nanocomposite. *Journal of Hazardous Materials*, *186*, 150–159.

Bhaumik, M., Maity, A., Srinivasu, V., Onyango, M.S. (2011b). Enhanced removal of Cr(VI) from aqueous solution using polypyrrole/Fe_3O_4 magnetic nanocomposite. *Journal of Hazardous Materials*, *190*, 381–390.

Bhaumik, M., Maity, A., Srinivasu, V., Onyango, M. S. (2012). Removal of hexavalent chromium from aqueous solution using polypyrrole-polyaniline nanofibers. *Chemical Engineering Journal*, *181*, 323–333.

Chauhan, G.S., Singh, B., Chauhan, S., Verma, M., Mahajan, S. (2005). Sorption of some metal ions on cellulosic-based hydrogels. *Desalination*, *181(1–3)*, 217–224.

Chauhan, K., Chauhan, G.S., Ahn, J.-H. (2010). Novel polycarboxylated starch-based sorbents for Cu^{2+} ions. *ACS Industrial Engineering Chemical Research*, *49*(6), 2548–2556.

Checkol, F., Elfwing, A., Greczynski, G., Mehretie, S., Inganas, O., Admassie, S. (2018). Highly stable and efficient lignin-PEDOT/PSS composites for removal of toxic metals. *Advanced Sustainable Systems*, *2*(1), 1700114.

Chen, J.F., Wang, G.Q., Zeng, X.F., Zhao, H.Y., Cao, D.P., Yun, J., Tan, C.K. (2004). Toughening of polypropylene-ethylene copolymer with nanosized $CaCO_3$ and styrenebutadiene-styrene. *Journal of Applied Polymer Science*, *94*(2) 796–802.

Chen, M., Jiang, W., Wang, F., Shen, P., Ma, P., Gu, J., Mao, J., Li, F. (2013). Synthesis of highly hydrophobic floating magnetic polymer nanocomposites for the removal of oils from water surface. *Applied Surface Science*, *286*, 249–256.

Costa, R.G.F., Ribeiro, C., Mattoso, L.H.C. (2010). Morphological and photocatalytic properties of PVA/TiO_2 nanocomposite fibers produced by electrospinning. *Journal of Nanoscience Nanotechnology*, *10*(8) 5144–5152.

Cumbal, L., Sengupta, A.K. (2005). Arsenic removal using polymer-supported hydrated iron(III) oxide nanoparticles: Role of Donnan membrane effect. *Environmental Science Technology*, *39*, 6508–6515.

Daneshvar, N., Salari, D., Khataee, A.R. (2004). Photocatalytic degradation of azo dye acid red 14 in water on ZnO as an alternative catalyst to TiO_2. *Journal of Photochemistry and Photobiology A: Chemistry*, *162*, 317–322.

Datta, A., Priyam, A., Bhattacharyya, S.N., Mukherje, K.K., Saha, A. (2008). Temperature tunability of size in CdS nanoparticles and size dependent photocatalytic degradation of nitroaromatics. *Journal of Colloid Interface Science*, *322*(1), 128–135.

DeMarco, M.J., SenGupta, A.K., Greenleaf, J.E. (2003). Arsenic removal using a polymeric/inorganic hybrid sorbent. *Water Research*, *37*, 164–176.

Dong, T.T., Luo, H.J., Wang, Y.P., Hu, B.J., Chen, H. (2011). Stabilization of Fe–Pd bimetallic nanoparticles with sodium carboxymethyl cellulose for catalytic reduction of para-nitrochlorobenzene in water. *Desalination*, *271*, 11–19.

Friedrich, K., Fakirov, S., Zhang, Z. (Eds). (2005). *Polymer composites: From nano-to-macro-scale*. Springer, New York.

Gao, B., Gao, Y., Li, Y. (2010). Preparation and chelation adsorption property of composite chelating material poly(amidoxime)/SiO_2 towards heavy metal ions. *Chemical Engineering Journal*, 1168, 542–549.

Ghorbani, M., Eisazadeh, H. (2013). Removal of COD, color, anions and heavy metals from cotton textile wastewater by using polyaniline and polypyrrole nanocomposites coated on rice husk ash. Composite Part B Engineering, *45*, 1–7.

Hojamberdiev, M., Prasad, R.M., Morita, K., Schiavon, M.A., Riede, R. (2012). Polymer-derived mesoporous SiOC/ZnO nanocomposite for the purification of water contaminated with organic dyes. *Microporous and Mesoporous Materials*, 151, 330–338

Hua, M., Xiao, L., Pan, B., Zhang, Q. (2013). Validation of polymer-based nano-iron oxide in further phosphorus removal from bioeffluent: Laboratory and scaled up study. *Frontiers of Environmental Science & Engineering*, *7*(3), 435.

Huang, Q., Liu, M., Zhao, J., Chen, J., Zeng, G., Huang, H., Tian, J., Wen, Y., Zhang, X., Wei, Y. (2018). Facile preparation of polyethylenimine-tannins coated SiO_2 hybrid materials for Cu^{2+} removal. *Applied Surface Science*, 427, 535-544.

Huang, Q., Shi, X., Pinto, R.A. (2008). Tunable synthesis and immobilization of zero-valent iron nanoparticles for environmental applications. *Environment Science & Technology*, *42*(23), 8884–8889.

Hunne, M.A., Rojas, O.J., Lucia, L.A., Sain, M. (2008). Cellulosic nanocomposites: A review. *Bioresources*, *3*(3) 929–980.

Iketani, K., Sun, R.D., Toki, M., Hirota, K., Yamaguchi O. (2003). Sol–gel-derived TiO_2/poly(dimethylsiloxane) hybrid films and their photocatalytic activities, *Journal of Physical Chemistry Solid*, *64*(3), 507–513.

Jaiswal, M., Chauhan, D., Sakararamakrishnan, N. (2012). Copper chitosan nanocomposite: Synthesis, characterization, and application in removal of organophosphorous pesticide from agricultural runoff. *Environmental Science and Pollution Research*, *19*(6), 2055–2062.

Jankeviciute, A., Károly, Z., Tarakina, N.V., Szépvölgyi, J., Kareiva, A. (2013). Synthesis and characterization of spherical amorphous alumosilicate nanoparticles using RF thermal plasma method. *Journal of Non-Crystaline Solids*, *359*, 9–14.

Javadian, J., Sorkhrodi, F.Z., Koutenaei, B.B. (2014). Experimental investigation on enhancing aqueous cadmium removal via nanostructure composite of modified hexagonal type mesoporous silica with polyaniline/polypyrrole nanoparticles. *Journal of Industrial Engineering Chemistry*, *20*, 3678–3688.

Jiang, Z.M., Lv, L., Zhang, W.M., Du, Q., Pan, B.C., Yang, L., Zhang, Q.X (2011). Nitrate reduction using nanosized zero-valent iron supported by polystyrene resins: Role of surface functional groups. *Water Research.*, *45*, 2191–2198.

Jimenez, G., Ogata, N., Kawai, H., Takashi, O. (1997). Structure and thermal/mechanical properties of poly(e-caprolactone)-clay blend. *Journal of Applied Polymer Science*, *64*, 2211–2220.

Jin, W., Maduraiveeran, G. (2017). Electrochemical detection of chemical pollutants based on gold nanomaterials. *Trends Environmental Analytical Chemistry*, *14*, 28–36

Katsoyiannis, I.A., Zouboulis, A.I. (2002). Removal of arsenic from contaminated water sources by sorption onto iron-oxide-coated polymeric materials. *Water Research*, *36*, 5141–5155.

Kawasumi, M., Hasegawa, N., Kato, M., Usuki, A., Okada, A. (1997). Preparation and mechanical properties of polypropylene-clay hybrids. *Macromolecules*, *30*, 6333–6338.

Keledi, G., Hariab, J., Pukanszky, B. (2012). Polymer nanocomposites: Structure, interaction, and functionality. *Nanoscale*, *4*, 1919–1938.

Khaydarov, R.A., Khaydarov, R.R., Gapurova, O. (2010). Water purification from metal ions using carbon nanoparticle-conjugated polymer nanocomposites. Water Research, 44, 1927–1933.

Kim, H., Hong, H.J., Lee, Y.J., Shin, H.J. (2008). Degradation of trichloroethylene by zerovalent iron immobilized in cationic exchange membrane. Desalination, 223(1–3), 212–220.

Konstantinou, I.K., Sakellarides, T.M., Sakkas, V.A., Albanis, T.A. (2001). Photocatalytic degradation of selected s-triazine herbicides and organophosphorus insecticides over aqueous TiO_2 suspensions. *Environment Science & Technology*, *316*(2), 398–405.

Konstantinou, I.K., Albanis, T.A. (2004). TiO_2-assisted photocatalytic degradation of azo dyes in aqueous solution: Kinetic and mechanistic investigations: A review. *Applied Catalysis: B Environment*, *49*(1), 1–14.

Kubacka, A., Ferrer, M., Cerrada, M.L., Serrano, C., Sánchez-Chaves, M., Fernández-García, M., de Andrés, A., Riobóo, R.J.J., Fernández-Martín, F., Fernández-García, M. (2009). Boosting TiO_2-anatase antimicrobial activity: Polymer-oxide thin films. *Applied Catalysis B: Environmental*, *89*, 441–447.

Kumar, R., Kumar, A., Chauhan, K., Gupta, R., Ahn, J.-H., Chauhan, G.S. (2009). Removal of As(V) from water by pectin based active hydrogels following geochemical approach. *Bioresource Technology*, *100*, 1474–1477.

Kwon, Y.T., Song, K.Y., Lee, W.I., Choi, G.J., Do, Y.R. (2000). Photocatalytic behavior of WO_3-loaded TiO_2 in an oxidation reaction. *Journal of Catalysis*, *191*(1), 192–199.

Li, J., Wang, X., Zhao, G., Chen, C., Chai, Z., Alsaedi, A., Hayat, T., Wang, X. (2018). Metal–organic framework-based materials: Superior adsorbents for the capture of toxic and radioactive metal ions. *Chemical Society Review*, *47*, 2322–2356.

Li, Y., Wang, Q., Liu, F., Zhou, Z., Huang, X., Wang, N., Wang, T., Hao, H. (2024). $HZrO_2$@PPy nanocomposite for effective fluoride removal from environmental water. *Journal of Environmental Chemical Engineering*, *12*(16), 113339.

Lin, C.J., Liou, S.L., Lo, Y.H. (2005). Degradation of aqueous carbon tetrachloride by nanoscale zerovalent copper on a cation resin. *Chemosphere*, *59*, 1299–1307.

Lin, C.J., Liou, Y.H., Lo, S.L. (2009). Supported Pd/Sn bimetallic nanoparticles for reductive dechlorination of aqueous trichloroethylene. *Chemosphere*, *74*(2) 314–319.

Liu, Y., Liu, Z., Gao, J., Dai, J., Han, J., Wang, Y., Xie, J., Yan, Y. (2011). Selective adsorption behavior of Pb(II) by mesoporous silica SBA-116-supported Pb(II)-imprinted polymer based on surface molecularly imprinting technique. *Journal of Hazardous Materials*, *186*, 197–205.

Liu, Z., Zhang, F. (2010). Nano-zerovalent iron contained porous carbons developed from waste biomass for the adsorption and dechlorination of PCBs. *Bioresource Technology*, *101*(7), 2562–2564.

Lofrano, G., Carotenuto, M., Libralato, G., Domingos, R.F., Markus, A., Dini, L., Gautam, R.K., Baldantoni, D., Rossi, M., Sharma, S.K., Chattopadhyaya, M.C., Giugni, M., Meric, S. (2016). Polymer functionalized nanocomposites for metals removal from water and wastewater: An overview. *Water Research*, *92*, 22–37.

Lu, G., Li, X., Qu, Z., Zhao, Q., Li, H., Shen, Y., Chen, G. (2010). Correlations of WO_3 species and structure with the catalytic performance of the selective oxidation of cyclopentene to glutaraldehyde on WO_3/TiO_2 catalysts. *Chemical Engineering Journal, 159*(1–3), 242–246.

Mahdavian, A.R., Mirrahimi, M.A.-S. (2010). Efficient separation of heavy metal cations by anchoring polyacrylic acid on superparamagnetic magnetite nanoparticles through surface modification. *Chemical Engineering Journal*, 159, 264–271.

Mahfuz, H.A., Adnan, A., Rangari, V., Jeelani, S. (2005). Manufacturing and characterization of carbon nanotube/polyethylene composites. *International Journal of Nanoscience, 4*, 55–72.

Mbhele, Z.H., Salemane, M.G., van Sittert, C.G.C.E., Nedeljkovi, J.M., Djokovi, V., Luyt, A.S. (2003). Fabrication and characterization of silver-polyvinyl alcohol nanocomposites. *Chemistry of Materials, 15*, 5019–5024.

Mehwish, N., Kausar, A., Siddiq, M. (2014). Advances in polymer-based nanostructured membranes for water treatment. *Polymer Plastics Technology and Engineering, 53*, 1290–1316.

Meric, S. (2016). Polymer functionalized nanocomposites for metals removal from water and wastewater: An overview. *Water Research, 92*, 22–37.

Mohan, S., Singh, D.K., Kumar, V., Hasan, S.H. (2017). Effective removal of fluoride ions by rGO/ZrO_2 nanocomposite from aqueous solution: Fixed bed column adsorption modelling and its adsorption mechanism. *Journal of Fluorine Chemistry*, 194, 40–50.

Nasir, J.A., Rehman, Z.U., Shah, S.N.A., Khan, A., Butler, I.S., Catlow, C.R.A. (2020). Recent developments and perspectives in CdS-based photocatalysts for water splitting. *Journal of Materials Chemistry* A, *8*, 20752–20780.

Pan, B., Pan, B., Chen, X., Zhang, W., Zhang, X., Zhang, Q., Zhang, Q., Chen, J. (2006), Preparation and preliminary assessment of polymer-supported zirconium phosphate for selective lead removal from contaminated water. *Water Research, 40*, 2938–2946

Pan, B.J., Wu, J., Pan, B.C., Lv, L., Zhang, W.M., Xiao, L., Wang, X., Tao, X., Zheng, S. (2009). Development of polymer-based nanosized hydrated ferric oxides (HFOs) for enhanced phosphate removal from waste effluents. *Water Research, 43*(17) 4421–4429.

Ponder, S.M., Darab, J.G., Mallouk, T.E. (2000). Remediation of Cr(VI) and Pb(II) aqueous solutions using supported, nanoscale zero-valent iron. *Environmental Science & Technology, 34*(12), 2564–2569.

Qu, X., Alvarez, P.J.J., Li, Q. (2013). Applications of nanotechnology in water and wastewater treatment. *Water Research, 47*, 3931–3946.

Raman, N., Sudharsan, S., Pothiraj, K. (2012). Synthesis and structural reactivity of inorganic–organic hybrid nanocomposites—a review. *Journal of Saudi Chemical Society, 16*, 339–352.

Saeki, D., Nagashima, Y., Sawada, I., Matsuyama, H. (2016). Effect of hydrophobicity of polymer materials used for water purification membranes on biofilm formation dynamics. *Colloids Surface A: Physicochemical Engineering Aspects*, 506, 622–628.

Saleh, T.A. (2015a). Isotherm, kinetic, and thermodynamic studies on Hg(II) adsorption from aqueous solution by silica- multiwall carbon nanotubes. *Environment Science Pollution Research, 22*(21), 16721–16731.

Saleh, T.A. (2015b). Mercury sorption by silica/carbon nanotubes and silica/activated carbon: A comparison study. *Journal of Water Supply: Research Technology – AQUA, 64*(8), 892–903.

Saleh, T.A. (2018). Simultaneous adsorptive desulfurization of diesel fuel over bimetallic nanoparticles loaded on activated carbon. *Journal of Cleaner Production, 172*, 2123–2132

Saleh, T.A., Parthasarathy, P., Irfan, M. (2019). Advanced functional polymer nanocomposites and their use in water ultra-purification. *Trends in Environmental Analytical Chemistry, 24*, e00067.

Sarkar, S., Chatterjee, P.K., Cumbal, L.H., SenGupta, A.K. (2011). Hybrid ion exchanger supported nanocomposites: Sorption and sensing for environmental applications. *Chemical Engineering Journal*, *166*, 923–931.

Setshedi, K.Z., Bhaumik, M., Songwane, S., Onyango, M.S., Maity, A. (2013). Exfoliated polypyrrole-organically modified montmorillonite clay nanocomposite as a potential adsorbent for Cr(VI) removal. *Chemical Engineering Journal*, *222*, 186–197.

Sharma, G., Pathania, D., Naushad, M., Kothiyal, N. (2014). Fabrication, characterization and antimicrobial activity of polyaniline Th(IV) tungstomolybdophosphate nanocomposite material: Efficient removal of toxic metal ions from water. *Chemical Engineering Journal*, *251*, 413–421.

Shu, H.Y., Chang, M.C., Chen, C.C. (2010). Using resin supported nano zero-valent iron particles for decoloration of Acid Blue 113 azo dye solution. *Journal of Hazardous Materials*, *184*(1–3), 499–505.

Shukla, S., Kim, K.T., Baev, A., Yoon, Y.K. (2010). Fabrication and characterization of gold polymer nanocomposite plasmonic nanoarrays in a porous alumina template. *ACS Nano* *4*(4), 2249–2255.

Song, H., Carraway, E.R., Kim, Y.H., Batchelor, B., Jeon, B.H., Kim, J. (2008). Amendment of hydroxyapatite in reduction of tetrachloroethylene by zero-valent zinc: Its rate enhancing effect and removal of Zn(II). *Chemosphere*, *73*(9), 1420–1427.

Sylvester, P., Westerhoff, P., Möller, T., Badruzzaman, M., Boyd, O. (2007). A hybrid sorbent utilizing nanoparticles of hydrous iron oxide for arsenic removal from drinking water. *Environmental Engineering Science*, *24*, 104–112.

Takafuji, M., Ide, S., Ihara, H., Xu, Z. (2004). Preparation of poly (1-vinylimidazole)-grafted magnetic nanoparticles and their application for removal of metal ions. *Chemistry of Materials*, *16*, 1979–1983.

Tarannum, N., Kumar, D., Kumar, N. (2022). β-Cyclodextrin-based nanocomposite derivatives: State of the art in synthesis, characterization and application in molecular recognition. *Chemistry Select*, *7*, e202200140.

Thakur, S., Kumari, S., Dogra, P., Chauhan, G.S. (2014). A new guar gum-based adsorbent for the removal of Hg(II) from its aqueous solutions. *Carbohydrate Polymers*, *106*, 276–282.

Tong, M., Yuan, S., Long, H., Zheng, M. (2011). Reduction of nitrobenzene in groundwater by iron nanoparticles immobilized in PEG/nylon membrane. *Journal of Contaminant Hydrology*, *122*(1–4), 16–25.

Unnikrishnan, V., Zabihi, O., Ahmadi, M., Li, Q., Blanchard, P., Kiziltas, A., Naebe, M. (2021). Metal–organic framework structure–property relationships for high-performance multifunctional polymer nanocomposite applications. *Journal of Materials Chemistry* A, *9*, 4348–4378.

Varanasi, P., Fullan, A., Sidhu, S. (2007). Remediation of PCB contaminated soils using iron nanoparticles. *Chemosphere*, *66*(6) 1031–1038.

Venkateswarlu, S., Yoon, M. (2015). Core–shell ferromagnetic nanorod based on amine polymer composite (Fe_3O_4@DAPF) for fast removal of Pb(II) from aqueous solutions. *ACS Applied Materials Interfaces*, *7*(45), 25362–25372.

Vivani, R., Alberti, G., Costantino, F., Nocchetti, M. (2008). New advances in zirconium phosphate and phosphonate chemistry: Structural archetypes. *Microporous Mesoporous Materials*, *107*, 58–70.

Wang, D.S., Zhang, J., Luo, Q., Li, X.Y., Duan, Y., An, J. (2009). Characterization and photocatalytic activity of poly(3-hexylthiophene)-modified TiO_2 for degradation of methyl orange under visible light. *Journal of Hazardous Materials*, *169*(1–3), 546–550.

Wang, J., Pan, K., He, Q., Cao, B. (2013). Polyacrylonitrile/polypyrrole core/shell nanofiber mat for the removal of hexavalent chromium from aqueous solution. *Journal of Hazardous Materials*, *244*, 121–129.

Wang, X., Chen, C., Liu, H., Ma, J. (2008a). Preparation and characterization of PAA/PVDF membrane-immobilized Pd/Fe nanoparticles for dechlorination of trichloroacetic acid. *Water Research*, *42*(18), 4656–4664.

Wang, X., Pan, S., Zhang, M., Qi, J., Sun, X., Gu, C., Wang, L., Li, J. (2019). Modified hydrous zirconium oxide/PAN nanofibers for efficient defluoridation from groundwater. *Science of The Total Environment*, *685*, 401–409.

Wang, Z., Huang, W., Fennell, D.E., Peng, P. (2008b). Kinetics of reductive dichlorination of 1,2,3,4–TCDD in the presence of zero-valent zinc. Chemosphere, *71*(2) 360–368.

Wang, Z., Lu, Y., Liu, J., Dang, Z., Zhang, L., Wang, W. (2011). Preparation of nano-zinc oxide/EPDM composites with both good thermal conductivity and mechanical properties. *Journal of Applied Polymer Science*, *119*(2), 1144–1155.

Wu, M.-T., Tsai, Y.-L., Chiu, C.-W., Cheng, C.-C. (2016). Synthesis, characterization, and highly acid-resistant properties of crosslinking β-chitosan with polyamines for heavy metal ion adsorption. RSC Advances, *6*, 104754–104762.

Wu, S.J., Liou, T.H., Mi, F.L. (2009). Synthesis of zero-valent copper-chitosan nanocomposites and their application for treatment of hexavalent chromium. *Bioresource Technology*, *100*(19), 4348–4353.

Wu, S., Li, F., Wang, H., Fu, L., Zhang, B., Li, G. (2010). Effects of poly (vinyl alcohol) (PVA) content on preparation of novel thiol-functionalized mesoporous PVA/SiO_2 composite nanofiber membranes and their application for adsorption of heavy metal ions from aqueous solution. *Polymer*, *51*, 6203–6211.

Wu, W., He, T., Chen, J.F., Zhang, X., Chen, Y. (2006). Study on *in situ* preparation of nano calcium carbonate/PMMA composite particles. *Material Letters*, *60*(19),2410–24116.

Xiong, Z., Zhao, D., Pan, G. (2007). Rapid and complete destruction of perchlorate in water and ion-exchange brine using stabilized zero-valent iron nanoparticles. *Water Research*, *41*(15), 3497–3505.

Xu, X., Wang, Q., Choi, H.C. (2010). Encapsulation of iron nanoparticles with PVP nanofibrous membranes to maintain their catalytic activity. *Journal of Membrane Science*, *348*, 231–237.

Yavuz, C.T., Mayo, J.T., Yu, W.W., Prakash, A., Falkner, J. C., Yean, S., Cong, L., Shipley, H. J., Kan, A., Tomson, M., Natelson, D., Colvin, V.L. (2006). Low-field magnetic separation of monodisperse Fe_3O_4 nanocrystals. *Science*, *314*, 964–967

Yin, P., Xu, Q., Qu, R., Zhao, G., Sun, Y. (2010). Adsorption of transition metal ions from aqueous solutions onto a novel silica gel matrix inorganic–organic composite material. *Journal of Hazardous Materials*. *173*, 710–716.

Zabihi, O., Mojtaba Ahmadi, M., Nikafshar, S., Preyeswary, K.C., Naebe, M. (2018). A technical review on epoxy-clay nanocomposites: Structure, properties, and their applications in fiber reinforced composites. Composites Part B: Engineering, 135, 1–24.

Zabihi, O., Ahmadi, M., Liu, C., Mahmoodi, R., Li, Q., Naebe, M. (2020). Development of a low cost and green microwave assisted approach towards the circular carbon fibre composites. Composites Part B: Engineering, 184,107750.

Zare, E.N., Motahari, A., Sillanpää, M. (2018). Nanoadsorbents based on conducting polymer nanocomposites with main focus on polyaniline and its derivatives for removal of heavy metal ions/dyes: A review. *Environmental Research*, *162*, 173–195.

Zhang, K., Kemp, K.C., Chandra, V. (2012a). Homogeneous anchoring of TiO_2 nanoparticles on graphene sheets for waste water treatment. *Material Letters*, *81*, 127–130.

Zhang, Q., Pan, B., Chen, X., Zhang, W., Pan, B., Zhang, Q., Lv, L., Zhao, X. (2008). Preparation of polymer-supported hydrated ferric oxide based on Donnan membrane effect and its application for arsenic removal. *Science China Series B*, *51*, 379–385.

Zhang, Q., Pan, B., Zhang, S., Wang, J., Zhang, W., Lv, L. (2011). New Insights into nanocomposite adsorbents for water treatment: A case study of polystyrene-supported zirconium phosphate nanoparticles for lead removal. *Journal of Nanoparticle Research*, *13*, 5355–5364.

Zhang, S., Chen, Q., Jing, D., Wang, Y., Gao, L. (2012b). Visible photoactivity and anti-photocorrosion performance of PdS–CdS photocatalysts modified by polyaniline. *International Journal of Hydrogen Energy*, *37*, 791–796.

Zhang, W., Shi, X., Zhang, Y., Gu, W., Li, B., Xian, Y. (2013). Synthesis of water-soluble magnetic graphene nanocomposites for recyclable removal of heavy metal ions. *Journal of Material Chemistry* A, *1*, 1745–1753.

Zhao, X., Lv, L., Pan, B., Zhang, W., Zhang, S., and Zhang, Q. (2011). Polymer-supported nanocomposites for environmental application: A review. *Chemical Engineering Journal*, *170*, 381–394.

Zhao, G., Zhang, H., Fan, Q., Ren, X., Li, J., Chen, Y., Wang, X. (2010). Sorption of copper(II) onto super-adsorbent of bentonite–polyacrylamide composites. *Journal of Hazardous Materials*, *173*, 661–668.

Zhou, L., Wang, Y., Liu, Z., Huang, Q. (2009a). Characteristics of equilibrium, kinetics studies for adsorption of Hg(II), Cu(II), and Ni(II) ions by thiourea-modified magnetic chitosan microspheres. *Journal of Hazardous Materials*, *161*, 995–1002.

Zhou, Y.-T., Nie, H.-L., Branford-White, C., He, Z.-Y., Zhu, L.-M. (2009b). Removal of Cu^{2+} from aqueous solution by chitosan-coated magnetic nanoparticles modified with alpha-ketoglutaric acid. *Journal of Colloid Interface Science*, *330*, 29–37.

Zhu, H., Jiang, R., Xiao, L., Chang, Y., Guan, Y., Li, X., Zeng, G. (2009). Photocatalytic decolorization and degradation of Congo Red on innovative crosslinked chitosan/nano-CdS composite catalyst under visible light irradiation, *Journal of Hazardous Materials*, *169*(1–3), 933–940.

Zhu, Y., Ran, T., Li, Y., Guo, J., Li, W. (2006). Dependence of the cytotoxicity of multi-walled carbon nanotubes on the culture medium. *Nanotechnology*, *17*, 4668–4674.

Zouboulis, A.I., Katsoyiannis, I.A. (2002). Arsenic removal using iron oxide loaded alginate beads. *Industrial Engineering Chemistry Research, 41*, 6149–6155.

8 Environmental and Green Aspects of Polymer Nanocomposites

Alka Singh and Richa Tomar

8.1 INTRODUCTION

Nanotechnology involves manipulating matter at the atomic or molecular scale, typically between 1 and 100 nm (Darwish et al., 2022). Nanoparticles, with their nanoscale dimensions, exhibit unique physical and chemical properties, such as a high surface area to volume ratio and enhanced interfacial reactivity (Pourhashem et al., 2020; Carroccio et al., 2022). These properties enable nanoparticles to interact selectively with contaminants in water, air, and soil, offering significant potential for environmental technology advancements (Omanovi et al., 2020). Industrial development, urban growth, and population increases have escalated the demand for water, food, agricultural products, and other natural resources (Wróblewska-Krepsztul et al., 2018). Additionally, climate change, overuse, water supply depletion, and anthropogenic pollution have raised global concerns about the availability of clean water (Omanovi et al., 2020; Roshni et al., 2024). Water pollution, primarily from industrial, agricultural, and municipal wastewater discharges, poses significant public health challenges and threatens ecosystems (Potbhare et al., 2024; Vattikuti et al., 2024). Such pollution can lead to excessive nutrient loads and oxygen depletion in aquatic environments, endangering human health, wildlife, and aquatic life (Khodakarami and Bagheri, 2021).

Despite their advantages, nanoparticles face challenges such as mass transport issues, high-pressure drops in fixed-bed or flow-through systems, separation and reuse difficulties, and potential ecological and health risks due to environmental release (Agboola et al., 2021; Pachaiappan et al., 2021). Developing hybrid nanocomposites, where nanoparticles are impregnated or coated onto larger solid particles, offers a promising strategy to address these challenges (Pachaiappan et al., 2021). Nanocomposites are materials derived from nano-sized dimensions, offering enhanced properties that make them suitable replacements for conventional micro-composites (Saleh et al., 2020). Despite this potential, the preparation of nanocomposites presents challenges, similar to those faced during the early days of micro-composite development, primarily due to a lack of knowledge in this area. The efficiency of nanocomposites based on nanoparticles is influenced by properties such as matrix compatibility, dispersion, aspect ratio, and specific surface area, making them attractive for various applications. The interface between the polymer matrix

 DOI: 10.1201/9781003527602-8

and nanoparticles plays a crucial role. Achieving proper dispersion of nanoparticles is key to enhancing properties such as flame retardance, mechanical strength, barrier effectiveness, and thermal stability. Additionally, their high surface areas contribute significantly to improved material properties. Common materials for nanocomposite fabrication include carbon-based substances such as granular activated carbon, silica, cellulose, sands, and polymers (Pourhashem et al., 2020; Toto et al., 2022). Polymers are particularly favored for their adjustable pore space, surface chemistry, and robust mechanical strength, making them suitable for long-term use (Hiremath et al., 2021; Potbhare et al., 2024). Nanomaterials and polymers are leading the development of innovative materials for effective wastewater treatment (Islam et al., 2020). Their functional characteristics can be tailored to meet the diverse needs of various wastewater treatment technologies (Islam et al., 2020). Polymer nanocomposites (PNCs) represent an advanced class of materials that combine the beneficial properties of both polymers and nanomaterials to deliver superior performance (Chow and Ishak, 2020; Hassan et al., 2021; Potbhare et al., 2024). Understanding their fundamental properties and underlying science is essential for the rational development of these functional materials (Pachaiappan et al., 2021).

Polymer-based nanocomposites (PNCs) integrate the inherent properties of nanoparticles with the stability, processability, and enhancements offered by the polymer matrix (Lawal, 2020; Pourhashem et al., 2020). Typical nanoparticles used in PNCs include zero-valent metals, metallic oxides, biopolymers, and single-enzyme nanoparticles (SENs). These nanoparticles can be incorporated into porous resins, cellulose, carboxymethyl cellulose, chitosan, and alginate, among others (Glaskova-Kuzmina et al., 2021). The selection of polymeric supports is based on their mechanical and thermal properties, hydrophobic/hydrophilic balance, chemical stability, biocompatibility, and specific chemical functionalities such as solvation, wettability, and templating effects (Arunachalam, 2018; Islam et al., 2020; Abdelazeez et al., 2024).

Nanoparticles enhance base polymers, improving mechanical strength, modulus, stability, thermomechanical properties, permeability, thermal stability, flame retardancy, chemical resistance, appearance, weight, and conductivity (Bhat et al., 2021; Carroccio et al., 2022). Nanocomposites with nanoscale reinforcements outperform conventional composites due to superior interfacial adhesion (Arunachalam, 2018; Debasrita et al., 2021; Costa et al., 2022). Historically, synthetic materials like polyesters and nano clays were used, but their environmental and economic drawbacks have spurred interest in renewable polymer composites (Idumah and Obele, 2020; Mondal et al., 2020; Idumah, 2021; Darwish et al., 2022). Green composites, derived from biodegradable polymers and agricultural resources, offer sustainable solutions by reducing fossil fuel dependence, carbon emissions, and plastic waste, while enhancing manufacturing speed and recyclability (Ates et al., 2020; Torres-Giner, 2020; Al Harby et al., 2022). The growing environmental issues associated with the manufacturing and disposal of conventional polymers have driven the search for more eco-friendly alternatives. One promising solution is PNCs, which are polymers enhanced with nanoparticles. These materials exhibit distinct mechanical, thermal, and barrier properties.

This chapter explores the environmental benefits and sustainability of PNCs, emphasizing their synthesis methods, applications, and potential for future development.

8.2 CLASSIFICATION OF NANOCOMPOSITES BASED ON GREEN PROSPECTS

The classification of nanocomposites is contingent upon the characteristics of the reinforcement materials and the matrix substrate. As such, nanocomposites can be categorized into three classes based on the type of matrix material (Figure 8.1).

8.2.1 Polymer Matrix Nanocomposites

PNCs are composite materials consisting of a polymer matrix embedded with nano-sized additives such as nanotubes, fibers, clay, and spherical nanoparticles (Abdel Maksoud et al., 2020; Maguire et al., 2023). These nanocomposites exhibit enhanced properties including high elastic stiffness and strength. They also demonstrate excellent characteristics such as barrier resistance, magnetic efficiency, and electrical properties (Chowdhury et al., 2022). The preparation of PNCs typically involves dispersing fillers within a polymer matrix. The interaction between the polymer and nanofillers occurs through intermolecular forces and chemical bonding, which contribute to their strength and performance (Solomon et al., 2024; Vafaeva et al., 2024). Clay minerals are often used as filler materials to achieve high strength in these composites. Polymers are valued for their diverse properties such as lightweight nature, durability, corrosion resistance, and capabilities in gas barrier, heat resistance, and fire retardancy applications (Vafaeva et al., 2024). Industries widely accept these polymers due to their versatile characteristics. Various methods are employed for synthesizing PNCs, including melt blending, *in situ* polymerization, mixing of filler substrates, and other techniques such as polymer intercalation, direct mixing of polymer and fillers, and sol–gel methods (Al-Gethami et al., 2024; Gavande et al., 2024; Solomon et al., 2024; Vafaeva et al., 2024).

8.2.1.1 Types of Polymer Nanocomposites

Numerous types of polymers are utilized in the synthesis of PNCs. These include thermoplastics, thermosets, elastomers, and natural biodegradable polymers (Gavande et al., 2024; Vafaeva et al., 2024). Thermosets are particularly versatile in various fields. The performance of these polymers depends significantly on their

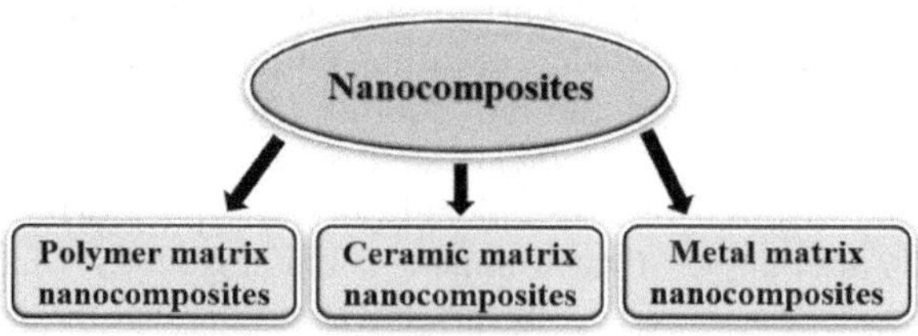

FIGURE 8.1 Classification of nanocomposites.

structure, including chemical composition and surface morphology. Thermosets form cross-links during curing, which are broken down at high temperatures (Gavande et al., 2024; Solomon et al., 2024; Vafaeva et al., 2024). They offer excellent mechanical properties and resistance to heat and chemicals in their environment. In contrast, thermoplastics undergo a reversible curing process without cross-linking, becoming soft at high temperatures and hardening upon cooling (Vafaeva et al., 2024).

8.2.1.2 The Emergence of Polymer Nanocomposites

PNCs emerged as a promising solution to enhance the properties of traditional polymers while addressing environmental concerns (Tajik et al., 2022; Solomon et al., 2024). The addition of nanoparticles such as clay, carbon nanotubes (CNTs), and metal oxides can significantly improve the mechanical, thermal, and chemical resistance properties of polymers (Al-Gethami et al., 2024; Gavande et al., 2024).

The key advantages of PNCs include:

i. Improved Properties: Enhanced strength, thermal stability, and barrier properties (Vafaeva et al., 2024) compared to traditional polymers.
ii. Material Efficiency: Smaller quantities of nanomaterials can achieve desired properties, reducing overall material usage (Carroccio et al., 2022).
iii. Potential for Biodegradability: When combined with biodegradable polymers, nanocomposites can be designed to degrade more rapidly than conventional plastics (Costa et al., 2022; Abdelazeez et al., 2024).

8.2.1.3 Biodegradable Polymer Nanocomposites

Biodegradable PNCs represent a promising category of materials combining biodegradability with enhanced properties from nanofillers (Debasrita et al., 2021; Gavande et al., 2024), thus reducing environmental plastic waste impact.

8.2.1.3.1 Polylactic Acid (PLA)-Based Nanocomposites

PLA, a widely used biodegradable polymer, benefits from nanoparticle reinforcement to enhance mechanical and thermal properties:

- PLA/Clay Nanocomposites: Clay nanoparticles improve barrier properties, making PLA suitable for packaging applications (Islam et al., 2020; Abdelazeez et al., 2024).
- PLA/Carbon Nanotube (CNT) Nanocomposites: CNTs enhance electrical conductivity and mechanical strength, expanding PLA's potential applications.

8.2.1.3.2 Polyhydroxyalkanoates (PHA)-Based Nanocomposites

Polyhydroxyalkanoates (PHAs), biodegradable polyesters from microbial fermentation, can be reinforced with nanoparticles to enhance properties:

- PHA/Cellulose Nanocomposites: Incorporating cellulose nanocrystals (CNCs) significantly boosts mechanical properties, suitable for packaging and biomedical devices (Wróblewska-Krepsztul et al., 2018; Idumah and Obele, 2020).

8.2.1.3.3 Starch-Based Nanocomposites

Starch, a natural polymer, can produce biodegradable nanocomposites:

- Starch/Nano clay Nanocomposites: Nano clay addition improves mechanical properties and water resistance, ideal for biodegradable packaging materials (Lawal et al., 2012; Darwish et al., 2022).

8.2.2 Ceramic Matrix Nanocomposites

Nanoceramic composites consist of a single phase at nanoscale dimensions, prized for their versatile electrical and mechanical properties, making them suitable for diverse applications (Al Harby et al., 2022; Abdelazeez et al., 2024). Various synthesis methods are employed, including conventional powder methods, polymer precursor routes, chemical processes, and techniques like spray pyrolysis such as the sol–gel method, colloidal processes, precipitation methods, and template approaches (Omanovi et al., 2020; Hassan et al., 2021). Popular examples of ceramic nanocomposites include Al_2O_3/SiO_2, SiO_2/Ni, Al_2O_3/TiO_2, and Al_2O_3/SiC. Additionally, CNTs are extensively integrated into ceramic nanocomposites such as Al_2O_3/CNT, $MgAl_2O_4$/CNT, and MgO/CNT, which find wide-ranging applications in the electronics industry (Omanovi et al., 2020; Hassan et al., 2021).

8.2.3 Metal Matrix Nanocomposites

Metal nanocomposites consist of substrates containing metal nanoparticles or alloy nanoparticles, offering enhanced physical, chemical, and mechanical properties compared to organic-phase nanocomposites (Zhao et al., 2011; Torres-Giner, 2020; Hassan et al., 2021; Khodakarami and Bagheri, 2021). These nanoparticles are employed to improve chemical resistance and mechanical strength. Recently, metal nanocomposites designed for high-temperature applications have been developed (Pachaiappan et al., 2021; Al Harby et al., 2022; Kumari et al., 2023). The nanoparticles act as barriers, impeding dislocation movement and thereby enhancing mechanical efficiency. Various synthesis methods have been employed, including liquid metal infiltration, spray pyrolysis, vapor deposition, rapid solidification, electrodeposition, and chemical methods like sol–gel and colloidal processes (Vignesh and Moorthi, 2017; Pachaiappan et al., 2021). Common metal nanocomposites include Fe-Cr/Al_2O_3, Ni/ Al_2O_3, Fe/MgO, Al/CNT, and Mg/CNT.

8.3 GREEN AND ENVIRONMENTAL ASPECTS IN THE SYNTHESIS OF POLYMER NANOCOMPOSITES

Green chemistry principles emphasize the use of environmentally benign methods for the synthesis of PNCs. The preparation of PNCs entails integrating diverse nanofillers into the polymer matrix (Lawal, 2020; Hassan et al., 2021). Various techniques, including *in situ* synthesis, solution mixing, melt processing, and electrospinning are employed for fabricating PNCs (Idumah and Obele, 2020; Torres-Giner,

2020; Darwish et al., 2022). The choice of preparation method hinges on parameters such as the specific polymeric system utilized, intended application domains, and criteria such as particle distribution and size (Gong et al., 2020; Islam et al., 2020; Abdelazeez et al., 2024). Essential strategies are given below:

1. Solvent-Free Processes: Developing synthesis routes that eliminate or minimize the use of harmful solvents.
2. Renewable Feedstocks: Utilizing renewable resources such as plant-based materials to produce polymers.
3. Energy-Efficient Methods: Employing processes that require less energy, such as microwave-assisted synthesis or ultrasonication.
4. Biocompatible Additives: Using nanoparticles derived from natural sources or those that are non-toxic and biocompatible.
5. Solvent-Free Processes

Solvent-free processes involve synthesizing PNCs without the use of organic solvents, which can be harmful to the environment and human health (Al-Gethami et al., 2024; Gavande et al., 2024). Techniques such as melt blending and in-situ polymerization are popular methods:

Melt Blending: This process involves melting the polymer and mixing it with nanoparticles at high temperatures. The absence of solvents makes it an environmentally friendly method (Arunachalam, 2018; Costa et al., 2022), although it requires precise temperature control to prevent degradation of the polymer.

In-Situ Polymerization: In this technique, monomers are polymerized in the presence of nanoparticles, leading to a uniform dispersion of the nanomaterials within the polymer matrix (Lawal et al., 2012; Wróblewska-Krepsztul et al., 2018; Darwish et al., 2022). The elimination of solvents reduces the risk of environmental contamination and simplifies the process (Figure 8.2).

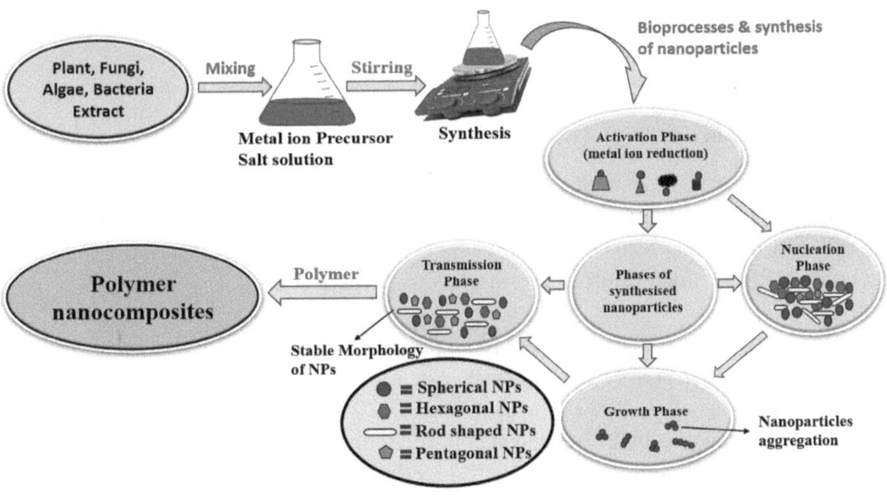

FIGURE 8.2 Schematic diagram of green synthesis of polymer nanocomposites.

An efficient and environmentally friendly approach to synthesizing nanoparticles should prioritize the use of renewable energy sources, minimize waste generation, and optimize energy consumption in accordance with green chemistry principles (Vignesh and Moorthi, 2017; Pachaiappan et al., 2021; Al-Gethami et al., 2024; Solomon et al., 2024). Green synthesis methods typically involve utilizing plants, microbes, biopolymers, and waste materials as bioactive components, coupled with low-energy processes and benign solvents (Carroccio et al., 2022; Gavande et al., 2024). Water, recognized as a universal solvent, is frequently employed in studies for its ability to extract bioactive polyphenols from various plant sources (Debasrita et al., 2021). These polyphenols serve as effective reducing, capping, and stabilizing agents during nanoparticle synthesis, sourced from leaves, stems, roots, flowers, fruit peels, seeds, gums, and agricultural residues (Al-Gethami et al., 2024; Vafaeva et al., 2024).

Agglomeration, a common issue in conventional chemical synthesis of nanoparticles, is notably reduced through green synthesis techniques (Bhat et al., 2021; Abdelazeez et al., 2024; Gavande et al., 2024). Plant extracts and microorganism-mediated processes leverage the polyphenolic content and natural proteins to stabilize and coat nanoparticles, thereby preventing agglomeration (Saleh et al., 2020; Debasrita et al., 2021). Nanoparticles prepared via green routes are typically stored in aqueous solutions at specific pH levels to ensure stability and uniform particle size (Gong et al., 2020). Additionally, enhancing stabilization involves using biocompatible reagents and chemical surface coatings that reduce reactivity upon exposure to environmental fluids (Bhat et al., 2021; Abdelazeez et al., 2024). For nanoparticles stored in powder form, maintaining a vacuum environment is crucial to prevent agglomeration caused by moisture from the surroundings (Arunachalam, 2018; Islam et al., 2020). The process entails in-situ polymerization where monomers disperse nanoparticles and undergo polymerization (Costa et al., 2022). The dispersed particles within the polymer matrix undergo additional polymerization reactions, facilitating composite development between intercalated sheets (Idumah and Obele, 2020; Mondal et al., 2020; Costa et al., 2022).

8.3.1 Renewable Feedstocks

Utilizing renewable feedstocks is pivotal in the synthesis of PNCs, aligning with principles of green chemistry (Sankar et al., 2022; Gavande et al., 2024; Vafaeva et al., 2024). Biomass-derived polymers such as polylactic acid (PLA), polyhydroxyalkanoates (PHA), and cellulose are increasingly integrated:

- Polylactic Acid (PLA): Derived from fermented plant starch (typically corn), PLA is biodegradable with a lower carbon footprint than petroleum-based plastics (Ates et al., 2020; Torres-Giner, 2020). Nanocomposites incorporating PLA can exhibit enhanced mechanical and thermal properties.
- Polyhydroxyalkanoates (PHA): Produced via microbial fermentation of sugars or lipids, PHAs are biodegradable and share properties akin to traditional

plastics (Torres-Giner, 2020; Hiremath et al., 2021). Nanoparticle reinforcement can further enhance their performance.

- Cellulose: As the most abundant natural polymer, cellulose is extractable from plants and utilized in producing nanocomposites (Thomas et al., 2019; Al-Gethami et al., 2024). CNCs and cellulose nanofibrils (CNFs) reinforce biodegradable polymers, resulting in materials possessing excellent mechanical properties (Gavande et al., 2024; Solomon et al., 2024).

8.3.2 Energy-Efficient Methods

Adopting energy-efficient synthesis methods is crucial for minimizing the environmental footprint of polymer nanocomposite production. The key methodologies include the following:

- Microwave-Assisted Synthesis: Utilizing microwaves enables rapid, uniform heating that accelerates synthesis processes, thereby reducing energy consumption (Al-Gethami et al., 2024). This method applies to both nanoparticle synthesis and nanocomposite preparation.
- Ultrasonication: High-frequency ultrasound waves facilitate efficient dispersion of nanoparticles within polymer matrices (Saleh et al., 2020), ensuring uniform distribution without the need for excessive heating, thus conserving energy.
- Electrospinning: By leveraging an electric field to draw polymer solutions into fine fibers, electrospinning enables the production of nanocomposite fibers with high surface area-to-volume ratios, suitable for various applications (Arunachalam, 2018; Islam et al., 2020; Abdelazeez et al., 2024).

8.3.3 Biocompatible Additives

The selection of biocompatible additives is pivotal in the development of environmentally friendly PNCs. Examples include:

- Chitosan: Derived from chitin found in crustacean exoskeletons, chitosan is biodegradable and non-toxic (Ibanez et al., 2018; Sankar et al., 2022). It enhances mechanical properties and biocompatibility when used as a reinforcing agent in nanocomposites.
- Starch: A natural polymer sourced from plants; starch contributes to the production of biodegradable nanocomposites (Pandiselvi and Thambidurai, 2014; Hosseini and Shahryari, 2017; Carroccio et al., 2022). Starch nanoparticles, for instance, improve the mechanical properties of polymers like PLA and PHA.
- Silk Fibroin: Extracted from silk, silk fibroin is a biocompatible protein-based biopolymer (Solomon et al., 2024; Vafaeva et al., 2024). It enhances the mechanical properties of biodegradable polymers, making them suitable for biomedical applications.

8.3.4 Sustainability Considerations

i. Renewable Resources: Utilizing biopolymers (e.g., polylactic acid, starch, cellulose) and natural nanofillers (e.g., cellulose nanocrystals, chitin nanofibers) derived from renewable sources reduces dependence on petroleum-based materials and promotes sustainability.
ii. Biodegradability: Developing PNCs that are biodegradable ensures that they break down into non-toxic byproducts, minimizing environmental impact. Biodegradable polymers combined with natural nanofillers enhance the composite's overall biodegradability.
iii. Energy Efficiency: Green PNCs often require less energy for production and processing compared to traditional composites. For instance, using lightweight nanocomposites can reduce energy consumption in transportation applications.
iv. Reduced Toxicity: Selecting non-toxic and eco-friendly materials for both the polymer matrix and the nanofillers helps in reducing the overall toxicity. This includes avoiding hazardous chemicals in the synthesis and processing stages.
v. Enhanced Properties: Nanofillers can significantly improve the mechanical, thermal, and barrier properties of biopolymers, making them more competitive with conventional plastics. This allows for the replacement of traditional materials with greener alternatives without sacrificing performance.
vi. Recycling and Reusability: Developing composites that are easier to recycle and repurpose can extend their lifecycle and reduce waste. Strategies include designing materials for disassembly and incorporating recycled content into new composites.
vii. Life Cycle Assessment (LCA): Conducting a comprehensive LCA to evaluate the environmental impact of PNCs from production to disposal helps in identifying areas for improvement and ensuring the overall sustainability of the materials.
viii. Applications in Green Technologies: Using PNCs in applications such as renewable energy (e.g., solar panels, wind turbines), water purification, and environmental remediation can contribute to broader sustainability goals.

8.4 APPLICATIONS OF GREEN POLYMER NANOCOMPOSITES

PNCs have found extensive application across diverse fields, as depicted in Figure 8.3.

8.4.1 Dye Removal

Dyes are extensively utilized across various industries such as textiles, paper production, printing inks, coatings, and cosmetics to impart color to products. However, these industries generate wastewater containing dyes that are known to be carcinogenic, toxic, and hazardous, posing significant environmental challenges (Al Harby et al., 2022). Therefore, effective removal of these dyes is crucial to render wastewater suitable for reuse in industrial processes and to mitigate environmental impacts.

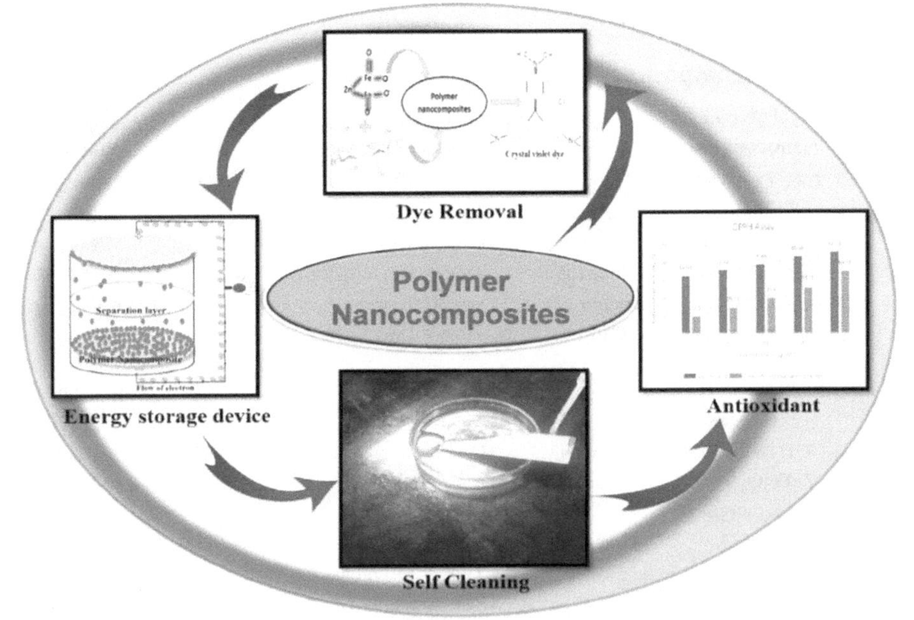

FIGURE 8.3 Applications of polymer nanocomposites.

PNCs have emerged as promising materials for the removal of dyes from contaminated water due to their unique properties. Table 8.1 presents several examples of PNCs that have been employed for the removal of dyes. PNCs combine the advantages of polymers and nanomaterials, offering large surface areas, numerous active sites, and robust chemical and thermal stability. For example, cellulose/zinc ferrite nanocomposites developed, which demonstrated exceptional efficiency in removing crystal violet dye through adsorption, achieving a removal rate of approximately 92% within 160 minutes with an adsorbent dose of 50 mg. Similarly, polyaniline (PANI)/zinc ferrite nanocomposites have been employed, achieving a maximum 90% removal with the same adsorbent dose for crystal violet dye removal in another study. The adsorption mechanisms of these PNCs involve chemical adsorption, intraparticle diffusion, as well as interactions such as hydrogen bonding, coordination, and electrostatic interactions, as shown in Figure 8.4.

These studies highlight the potential of PNCs as effective adsorbents for the removal of dyes from industrial wastewater, addressing both environmental concerns and the need for sustainable water management practices.

PNCs incorporating photocatalysts such as ZnO, TiO_2, and CuO nanoparticles have been effectively employed for dye removal. These photocatalysts generate hydroxyl free radicals capable of degrading organic dyes (Al Harby et al., 2022; Potbhare et al., 2024; Roshni et al., 2024; Vattikuti et al., 2024). Consequently, the combination of adsorption at the active sites of PNCs and photocatalytic degradation appears to be a highly effective method for dye removal. A study demonstrated the efficiency of chitosan (Ct) and chitosan/ZnO (Ct/ZnO) in removing methylene blue

TABLE 8.1
Polymer Nanocomposites in Dye Removal

S. No.	Polymer Nanocomposites	Dye	Dose of Nanocomposites (g/L)	Efficacy	References
1.	$ZnFe_2O_4$-PANI nanocomposite	Crystal violet (CV)	Dose = 0.05 g, Time= 160 minutes	90%	(Singh and Tomar, 2024)
2.	Chitosan/ZnO nanocomposite	Methylene blue (MB)	Dose = 0.015 g, Time = 180 minutes	96%	(Mostafa et al., 2020)
3.	$ZnFe_2O_4$@BC nanocomposite	Indigo Carmine	Dose = 0.01 g, Time = 180 minutes	69%	(Ajibade and Nnadozie, 2020)
4.	PANI@ZnO-SiO_2 nanocomposite	Congo Red	NA	83%	(Benchikh et al., 2022)
5.	GQDs@PEG@Mg $ZnFe_2O_4$ nanocomposite	Methylene blue (MB)	NA	98%	Perveen et al. (2023)
6.	PANI@WO_3 nanocomposite	Orange G	Dose = 0.225 g, Time = 180 minutes	90%	Hasanin et al. (2021)
7.	$CoFe_2O_4$ /MWCNT nanocomposite	Crystal violet (CV)	Dose = 15 g, Time = 180 minutes	96%	Tariq et al. (2023)
8.	Titanate nanotubes	Crystal violet (CV)	Dose = 3 g, Time = 1 hours	75%	Mohanty et al. (2020)
9.	Polypyrrole/Zeolite nanocomposite	Reactive Blue (RB) Reactive Red (RR)	Dose = 1.8 g, Time = 80 minutes	86% of RB 88% of RR	Senguttuvan et al. (2022)
10.	Molecularly imprinted Chitosan/TiO_2 nanocomposite	Rose Bengal (RB)	Dose = 0.02 g, Time = 60 minutes	85%	Ahmed et al. (2017)

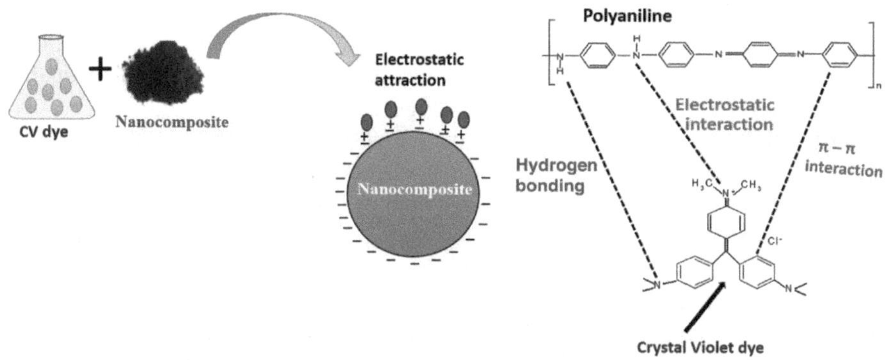

FIGURE 8.4 Schematic diagram of adsorption mechanism by $ZnFe_2O_4$/PANI nanocomposites (Singh and Tomar, 2024).

dye, reporting removal rates between 81% and 96.7%, respectively. This improvement was due to the synergistic effects of adsorption and the photocatalytic activity of ZnO nanoparticles. Similarly, Kumar et al. developed a Ct/CuO film for the removal of malachite green and methylene blue dyes, achieving high removal efficiencies of 99.02% for malachite green and 93.55% for methylene blue.

The performance of PNCs in dye removal can be optimized by controlling factors such as the types and properties of the polymers and nanomaterials used, as well as the methods of PNC preparation and design. A combination of adsorption and photocatalytic degradation mechanisms typically results in enhanced performance compared to using either mechanism alone. Furthermore, PNCs with magnetic properties are advantageous due to their ease of recovery and reusability.

Experimental conditions such as dye concentration, PNC concentration, pH, temperature, and contact time also influence the dye removal process. Integrating these factors can lead to the development of PNCs with exceptional dye removal capabilities. While PNCs have shown promise in laboratory settings, scaling up production for industrial applications remains a challenge, requiring further research to reduce production costs and improve practical applicability.

8.4.2 Energy Storage

Energy is a critical factor in the development of both developed and developing nations. As the global population increases and living standards improve, the demand for energy rises significantly (Toto et al., 2022; George et al., 2024). Currently, standard energy resources rely heavily on fossil fuels, which are not only finite but also contribute to pollution and global warming. Therefore, the development of renewable energy storage and conversion materials, along with their associated devices, is of paramount importance.

To address this energy crisis, numerous scientific research centers worldwide are receiving funding to explore new solutions and enhance existing energy technologies. The challenges can be tackled in two primary ways. First, energy can be efficiently converted from inexhaustible sources, such as through oxygen reduction reactions, solar power, and water, into usable forms like electricity and fuel. Key materials for this purpose include fuel cells, solar cells, and water-splitting catalysts.

Second, there is a need for economical, environmentally friendly, practical, and high-performance energy storage devices. Renewable energy sources largely meet these criteria. Among these, lithium-based batteries, such as lithium-ion, lithium-air, and lithium-sulfur batteries, are particularly promising for storage applications. Another significant and efficient energy storage device is the supercapacitor, which can store and release energy rapidly. For example, developed supercapacitor materials utilizing a ZnO-PANI nanocomposite integrated with porous carbon and redox-active electrolytes as shown in Figure 8.5. The resulting material demonstrated a specific capacitance (Csp) of 367 Fg^{-1} and retained 98% of its capacitance at a current density of 4 A/g, even after 800 cycles. This finding highlights the potential of the ZnO-PANI nanocomposite for use in supercapacitors.

Sajid Abbas developed a ternary nanocomposite consisting of reduced graphene oxide (rGO), PANI, and zinc oxide (ZnO) using a polymerization method (Abbas

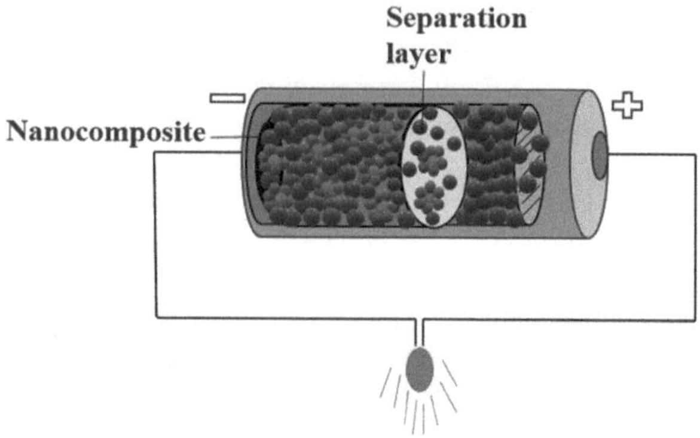

FIGURE 8.5 Working mechanism of symmetric supercapacitor.

et al., 2023). This nanocomposite was characterized using various analytical techniques. The findings indicate that the rGO/PANI/ZnO nanocomposite exhibits superior capacitive behavior, achieving a specific capacity of 1546 F/g at a scan rate of 2 mV/s on an eggshell membrane electrode (ESME). Additionally, the rGO/PANI/ZnO nanocomposite demonstrated excellent cycling stability, maintaining 97% of its capacity after 3,000 cycles. This high performance is attributed to the rapid electron transfer between the rGO/ZnO and PANI components in the energy storage device.

Table 8.2 provides a summary of energy storage applications utilizing nanocomposites derived from bio-renewable resources.

8.4.3 Antioxidant

According to researches, free radicals are a primary cause of oxidative stress and related health issues. One particularly harmful free radical, DPPH, has been shown to adversely affect human health. Soren et al. (2018) propose that the antioxidant properties of $ZnFe_2O_4$/cellulose nanocomposites are likely due to electron donation from the highly dense oxygen atom in the nanocomposite to the unpaired electron on the nitrogen atom. This interaction reduces the intensity of the $n \rightarrow \pi^*$ transition at 517 nm and eliminates the violet color characteristic of DPPH.

Richa et al. (2023) investigated the antioxidant activity of $ZnFe_2O_4$/cellulose nanocomposites compared to ascorbic acid, using the DPPH scavenging assay, as shown in Figure 8.6. The results revealed that the $ZnFe_2O_4$/cellulose nanocomposites demonstrated significantly lower antioxidant activity than ascorbic acid, with IC_{50} values of 49.64 and 41.60 μg/mL, respectively. These findings underscore the potential of $ZnFe_2O_4$/cellulose-based nanocomposites as effective natural antioxidants for protecting against oxidative stress related to degenerative diseases.

8.4.4 Self-Cleaning coatings

The application of coatings plays a crucial role in altering the surface properties of materials. By integrating thermoset polymer nano-coatings, enhancements in durability and wettability are achieved. Furthermore, these coatings enhance

TABLE 8.2
Polymer Nanocomposites in Energy Storage Devices

S. No.	Adsorbent	Electrolyte	Specific Capacitance	Cyclic Performance	Density	Reference
1.	CH-ZnO/PANI	1 M KOH	587.15 Fg^{-1}	1,000 (80%)	175 mA cm^{-2}	Pandiselvi and Thambidurai (2014)
2.	CH/GM/PANI	1 M KOH	609.2 Fg^{-1}	500 (96%)	5 Ag^{-1}	Hosseini and Shahryari (2017)
3.	ZnO/styrene maleic anhydride (SMA)	1 M KOH	268.5 Fg^{-1}	1,000 (80%)	0.1 Ag^{-1}	Chakraborty and Mary (2020)
4.	ZnO@Cfoam	3 M KOH	1120 m Fcm^{-2}	3,000 (95.8%)	4 mA cm^{-2}	Fahimi and Moradlou (2020)
5.	CH-graphene-PANI	2 M KOH	1519 Fg^{-1}	1,000 (88%)	0.1 Ag^{-1}	Torvi et al. (2018)
7.	PPY@ZrO_2-ZnO	1 M HCl	395.3 Fg^{-1}	4,000 (82.4%)	0.5 Ag^{-1}	Zenasni et al. (2023)
8.	Polyindole/ nickel oxide/ ZnO (PNZ)	1 M H_2SO_4	2345.8 Fg^{-1}	5,000 (78.5%)	100 mVs^{-1}	Humayun et al. (2023)
10.	PANI/GO/ZFO	1 M H_2SO_4	594 Fg^{-1}	1,500 (82%)	1 Ag^{-1}	Ghasemi et al. (2022)
11.	Polyaniline/CuO	1 M KCl	286.6 Fg^{-1}	2,000 (89%)	500 mV/s	Ashokkumar et al. (2019)
12.	Fe_2O_3/Polyaniline	1 M KCl	1033 Fg^{-1}	15,000 (82%)	5 mV/s	Senapati and Chakraborty (2021)

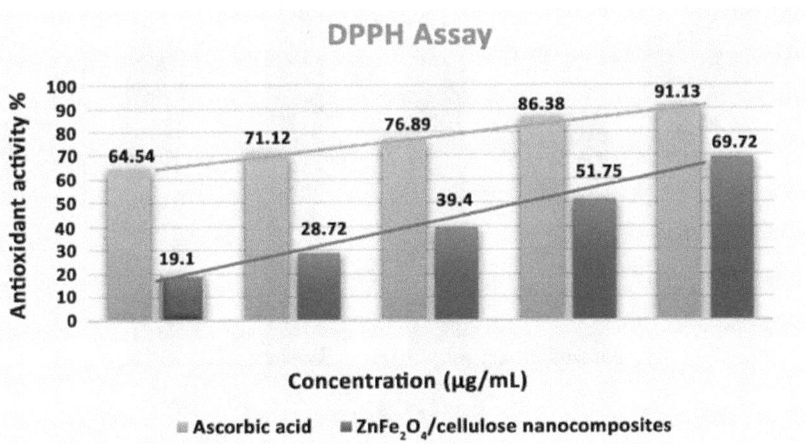

FIGURE 8.6 Antioxidant efficacy of ascorbic acid (positive control) and $ZnFe_2O_4$/Cellulose nanocomposites against DPPH (Singh et al., 2023).

chemical, weather, and resilience to abrasion properties. Recent advancements in nanocomposites have shown significant promise across various fields, particularly in pharmacology and biotechnology, as noted by biologists. However, the medical applicability of some hybrid materials is hindered by their intrinsic toxicity. Inorganic antimicrobial agents, known for their robust thermal stability and broad spectrum of microbial resistance, are increasingly utilized in antiviral and anticancer therapies, targeted drug delivery systems, and the development of antibacterial materials.

Hydrophobicity is a widespread phenomenon observed in numerous organisms. Numerous prominent examples in nature exhibit inherently hydrophobic characteristics, such as the surface of a lotus leaf, the legs of water striders, and the eyes of mosquitoes. Researchers transitioning from natural phenomena to biomimetic materials focus on two key factors for creating hydrophobic surfaces: integrating micro/nano hierarchical structures and incorporating chemical compositions with low surface energy. These hydrophobic surfaces, capable of self-cleaning, have found extensive use in practical applications.

For example, engineered a superhydrophobic aluminum (Al) surface using anodization, specifically designed to prevent icing at temperatures as low as 24°C, demonstrating exceptional self-cleaning properties. Similarly, utilized a spray coating technique to produce hydrophobic surfaces with high thermal stability, making them promising for outdoor self-cleaning applications. According to Sutar et al. (2020), the PVC/SiO_2 nanocomposite-coated surfaces displayed hydrophobic characteristics, evidenced by a water contact angle of 170° (Figure 8.7).

This high contact angle indicates a hydrophobic surface where water droplets maintain a spherical shape after different layers. Such surfaces are effective in self-cleaning applications, where spherical water droplets can easily remove dust

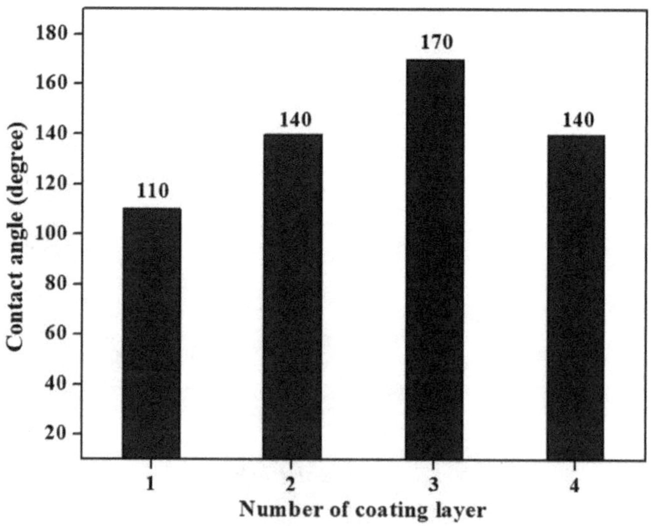

FIGURE 8.7 Contact angle measurement of a water droplet on a surface modified with a PVC/SiO_2 nanocomposite (Sutar et al., 2020).

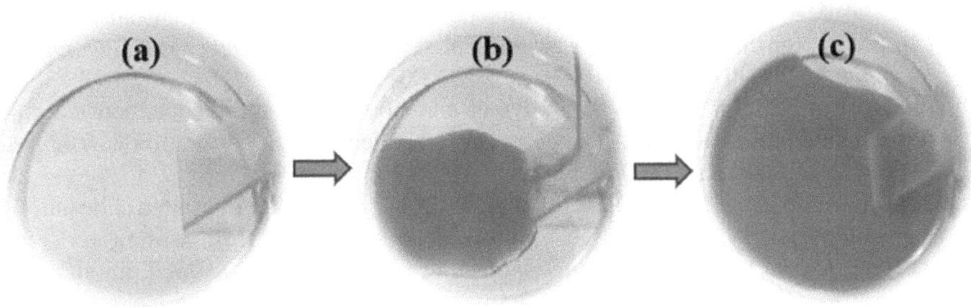

FIGURE 8.8 Image illustrates the self-cleaning characteristics of a superhydrophobic coating (a-c) (Sutar et al., 2020).

particles as they roll off. To demonstrate this, muddy water was used as a test material. When 50 mL of muddy water was applied to the hydrophobic coating, water droplets swiftly repelled it, preserving the surface's cleanliness (Figure 8.8a–c). These results underscore the strong water-repellent properties and excellent self-cleaning performance of the developed superhydrophobic coating.

8.6 CONCLUSION

This chapter aimed to elucidate the current state of PNCs across various applications. As highlighted, these applications necessitate the production of stable, functionalized nanoparticles with biocompatible properties. A major challenge for researchers in this field is the development of nanoparticles with specific functional attributes suitable for industrial and environmental applications. PNCs are pivotal to advancements in material science, offering promising physical and chemical properties that can extend smart behaviors to material applications. However, further research is needed to fully understand these systems. Mastery over the critical and complex variables is essential for advancing these materials and realizing their potential for scale-up. Various preparation techniques for PNCs have been documented, but additional research is required due to the ongoing challenges of ensuring uniform dispersion and optimizing the intrinsic properties of nanofillers. The preparation method and type of nanofillers are crucial in determining the appropriate application for a nanocomposite. PNCs show potential in applications such as water treatment, energy storage devices, antioxidants, and self-cleaning material, owing to their enhanced properties compared to the original polymers.

REFERENCES

Abbas, S. et al. (2023) 'Facile synthesis of rGO/PANI/ZnO ternary nanocomposites for energy storage devices', *Journal of the Korean Ceramic Society*, 60(1), pp. 127–140. doi:10.1007/s43207-022-00250-9.

Abdel Maksoud, M.I.A. et al. (2020) 'Insight on water remediation application using magnetic nanomaterials and biosorbents', *Coordination Chemistry Reviews*, 403. doi:10.1016/j.ccr.2019.213096.

Abdelazeez, A.A.A. et al. (2024) 'Polymer nanocomposites: catalysts for sustainable hydrogen production from challenging water sources', *Advanced Energy and Sustainability Research*, p. 2400077. doi:10.1002/aesr.202400077.

Agboola, O. et al. (2021) 'A review on polymer nanocomposites and their effective applications in membranes and adsorbents for water treatment and gas separation', *Membranes*, 11(2), pp. 1–33. doi:10.3390/membranes11020139.

Ahmed, M.A., Abdelbar, N.M. and Mohamed, A.A. (2017) 'Molecular imprinted chitosan-TiO2 nanocomposite for the selective removal of Rose Bengal from wastewater', *International Journal of Biological Macromolecules* [Preprint]. doi:10.1016/j.ijbiomac.2017.09.082.

Ajibade, P.A. and Nnadozie, E.C. (2020) 'Synthesis and structural studies of manganese ferrite and zinc ferrite nanocomposites and their use as photoadsorbents for indigo carmine and methylene blue dyes', *ACS Omega*, 5(50), pp. 32386–32394. doi:10.1021/acsomega.0c04404.

Al-Gethami, W. et al. (2024) 'Emerging environmentally friendly bio-based nanocomposites for the e ffi cient removal of dyes and micropollutants from wastewater by adsorption: a comprehensive review', *RSC Advances*, pp. 2804–2834. doi:10.1039/d3ra06501d.

Al Harby, N.F., El-Batouti, M. and Elewa, M.M. (2022) 'Prospects of polymeric nanocomposite membranes for water purification and scalability and their health and environmental impacts: a review', *Nanomaterials*, 12(20), p.3637.

Arunachalam, P. (2018) 'Polymer-based nanocomposites for energy and environmental applications', in *Polymer-Based Nanocomposites for Energy and Environmental Applications*. Woodhead Publishing, pp. 185–203. doi:10.1016/B978-0-08–102262-7.00006-4.

Ashokkumar, S.P. et al. (2019) 'Structure, morphology, thermal and electrochemical studies of electrochemically synthesized polyaniline/copper oxide nanocomposite for energy storage devices', *Materials Research Express*, 6(12). doi:10.1088/2053-1591/ab5dde.

Ates, B. et al. (2020) 'Chemistry, structures and advanced applications of nanocomposites from bio-renewable resources', *Chemical Reviews*, 120(17), pp. 9304–9362. doi:10.1021/acs.chemrev.9b00553.

Benchikh, I. et al. (2022) 'Development and characterization of novel hybrid materials formed from ZnO-SiO2-modified and polyaniline for organic dyes adsorption and electrochemical supercapacitor applications', *Research Square*, pp. 1–17. DOI: https://doi.org/10.21203/rs.3.rs-1479466/v1

Bhat, A. et al. (2021) 'Review on nanocomposites based on aerospace applications', *Nanotechnology Reviews*, 10(1), pp. 237–253.

Carroccio, S.C. et al. (2022) 'Impact of nanoparticles on the environmental sustainability of polymer nanocomposites based on bioplastics or recycled plastics – a review of life-cycle assessment studies', *Journal of Cleaner Production*, 335(April 2021), p. 130322. doi:10.1016/j.jclepro.2021.130322.

Chakraborty, S. and Mary, N.L. (2020) 'Biocompatible supercapacitor electrodes using green synthesised ZnO/Polymer nanocomposites for efficient energy storage applications', *Journal of Energy Storage*, 28(October 2019). doi:10.1016/j.est.2020.101275.

Chow, W.S. and Ishak, Z.A.M. (2020) 'Smart polymer nanocomposites: a review', *Express Polymer Letters*, 14(5), pp. 416–435.

Chowdhury, I.R. et al. (2022) 'Removal of lead ions (Pb2+) from water and wastewater: a review on the low-cost adsorbents', Applied Water Science. Springer International Publishing. doi:10.1007/s13201-022-01703-6.

Costa, J. et al. (2022) 'Kaolinite review: intercalation and production of polymer nanocomposites', *Engineered Science*, 17(11), pp. 28–44.

Darwish, M.S.A., Mostafa, M.H. and Al-Harbi, L.M. (2022) 'Polymeric nanocomposites for environmental and industrial applications', *International Journal of Molecular Sciences*, 23(3), p. 1023.

Debasrita, B. et al. (2021) 'A materials science approach towards bioinspired polymeric nanocomposites: a comprehensive review', *International Journal of Polymeric Materials and Polymeric Biomaterials*, 0(0), pp. 1–16. doi:10.1080/00914037.2021.1990057.

Fahimi, Z. and Moradlou, O. (2020) 'Fabrication of ZnO@C foam: a flexible free-standing electrode for energy storage devices', *Materials and Design*, 189, p. 108525. doi:10.1016/j.matdes.2020.108525.

Gavande, V. et al. (2024) 'A systematic review on green and natural polymeric nanofibers for biomedical applications', *International Journal of Biological Macromolecules*, 262(P2), p. 130135. doi:10.1016/j.ijbiomac.2024.130135.

George, N.S. et al. (2024) 'Hydrothermally grown hollandite manganese dioxide nanorods: evaluation of supercapattery performance and photocatalytic efficiency', *Ionics*. doi:10.1007/s11581-024-05616-z.

Ghasemi, A.K. et al. (2022) 'Controllable synthesis of zinc ferrite nanostructure with tunable morphology on polyaniline nanocomposite for supercapacitor application', *Journal of Energy Storage*, 51(April), p. 104579. doi:10.1016/j.est.2022.104579.

Glaskova-Kuzmina, T. et al. (2021) 'Durability of biodegradable polymer nanocomposites', *Polymers*, 13, pp. 3375.

Gong, M., Zhang, L. and Wan, P. (2020) 'Polymer nanocomposite meshes for flexible electronic devices', *Progress in Polymer Science*, p. 101279. doi:10.1016/j.progpolymsci.2020.101279.

Hasanin, M. et al. (2021) 'Photocatalytic degradation of pesticide intermediate using green eco-friendly amino functionalized cellulose nanocomposites', *Materials Science and Engineering B: Solid-State Materials for Advanced Technology*, 270(September 2020), p. 115231. doi:10.1016/j.mseb.2021.115231.

Hassan, T. et al. (2021) 'Functional nanocomposites and their potential applications: a review', *Journal of Polymer Research*, pp. 1–22. doi:10.1007/s10965-021-02408-1.

Hiremath, A. et al. (2021) 'Nanoparticles filled polymer nanocomposites: a technological review', *Cogent Engineering*, 8(1). doi:10.1080/23311916.2021.1991229.

Hosseini, M.G. and Shahryari, E. (2017) 'A novel high-performance supercapacitor based on chitosan/graphene oxide-MWCNT/polyaniline', *Journal of Colloid and Interface Science*. doi:10.1016/j.jcis.2017.02.027.

Humayun, H. et al. (2023) 'Polyindole embedded nickel/zinc oxide nanocomposites for high-performance energy storage applications', *Nanomaterials*, 13(3). doi:10.3390/nano13030618.

Ibanez, J.G. et al. (2018) 'Conducting polymers in the fields of energy, environmental remediation, and chemical – chiral sensors', *Chemical Reviews*. doi:10.1021/acs.chemrev.7b00482.

Idumah, C.I. (2021) 'Progress in polymer nanocomposites for bone regeneration and engineering', *Polymers and Polymer Composites*, 29(5), pp. 509–527. doi:10.1177/0967391120913658.

Idumah, C.I. and Obele, C.M. (2020) 'Understanding interfacial influence on properties of polymer nanocomposites', *Surfaces and Interfaces*, p. 100879. doi:10.1016/j.surfin.2020.100879.

Islam, A., Merca, T.D. and Doagou-Rad, S. (2020) 'An application-oriented roadmap to select polymeric nanocomposites for advanced applications', *Polymer Composites*. doi:10.1002/pc.25461.

Khodakarami, M. and Bagheri, M. (2021) 'Recent advances in synthesis and application of polymer nanocomposites for water and wastewater treatment', *Journal of Cleaner Production*, 296, p. 126404. doi:10.1016/j.jclepro.2021.126404.

Kumari, H. et al. (2023) 'A review on photocatalysis used for wastewater treatment: dye degradation', Water, Air, and Soil Pollution. Springer International Publishing. doi:10.1007/s11270-023-06359-9.

Lawal, A.T. (2020) 'Recent progress in graphene based polymer nanocomposites', *Cogent Chemistry*, 6(1). doi:10.1080/23312009.2020.1833476.

Lawal, G.I., Balogun, S.A. and Akpan, E.I. (2012) 'Review of green polymer nanocomposites', *Journal of Minerals and Materials Characterization and Engineering*, 11(4), pp. 385–416.

Maguire, J.M., Koutsos, V. and Ray, D. (2023) 'Vacuum-infused thermoplastic fibre-metal laminates – advances in bonding and recycling', *Materials Today Communications*, 37(July), p. 106961. doi:10.1016/j.mtcomm.2023.106961.

Mohanty, S., Moulick, S. and Maji, S.K. (2020) 'Adsorption/photodegradation of crystal violet (basic dye) from aqueous solution by hydrothermally synthesized titanate nanotube (TNT)', *Journal of Water Process Engineering*, 37(June), p. 101428. doi:10.1016/j.jwpe.2020.101428.

Mondal, P., Anweshan, A. and Purkait, M.K. (2020) 'Chemosphere green synthesis and environmental application of iron-based nanomaterials and nanocomposite: a review', *Chemosphere*, 259, p. 127509. doi:10.1016/j.chemosphere.2020.127509.

Mostafa, M.H. et al. (2020) 'Microwave-assisted preparation of Chitosan/ZnO nanocomposite and its application in dye removal', *Materials Chemistry and Physics*, 248(March), p. 122914. doi:10.1016/j.matchemphys.2020.122914.

Omanovi, E. et al. (2020) 'Nanocomposites: a brief review', *Health and Technology*, 10(1), pp. 51–59.

Pachaiappan, R. et al. (2021) 'Metal/metal oxide nanocomposites for bactericidal effect: a review', *Chemosphere*, 272, p. 128607. doi:10.1016/j.chemosphere.2020.128607.

Pandiselvi, K. and Thambidurai, S. (2014) 'Chitosan-ZnO / polyaniline ternary nanocomposite for high-performance supercapacitor', *Ionics*, pp. 551–561. doi:10.1007/s11581-013-1020-0.

Perveen, S. et al. (2023) 'The synergistic triad of graphene quantum dots, polymer, and ferrites for the photodegradation of dyes in aqueous solution', *Heliyon*, 9(11), p. e21739. doi:10.1016/j.heliyon.2023.e21739.

Potbhare, A.K. et al. (2024) 'Bioinspired graphene-based metal oxide nanocomposites for photocatalytic and electrochemical performances: an updated review', *Nanoscale Advances*, pp. 2539–2568. doi:10.1039/d3na01071f.

Pourhashem, S. et al. (2020) 'Inorganic nanocomposite coatings with superior corrosion protection performance: a review', *Journal of Industrial and Engineering Chemistry*, 88, pp. 29–57. doi:10.1016/j.jiec.2020.04.029.

Roshni, C.P. et al. (2024) 'Synthesis and characterization of alpha and beta cobalt hydroxide nanostructures for photocatalytic dye degradation and supercapacitor applications', *Next Materials*, 4(March), p. 100199. doi:10.1016/j.nxmate.2024.100199.

Saleh, T.A. et al. (2020) 'Recent trends in functionalized nanoparticles loaded polymeric composites: an energy application', *Materials Science for Energy Technologies*, 3, pp. 515–525. doi:10.1016/j.mset.2020.05.005.

Sankar, M.S.K. et al. (2022) 'The aspect of green nanocomposites in green technology and sustainable development: state of the art and new challenges', *Nanomaterials for Innovative Energy Systems and Devices*. Springer Nature Singapore. doi:10.1007/978–981-19–0553–7.

Senapati, A. and Chakraborty, A.K. (2021) 'Flowerlike Fe2O3–polyaniline nanocomposite as electrode for supercapacitor', *Journal of Materials Science: Materials in Electronics*, 32(23), pp. 27794–27800. doi:10.1007/s10854-021-07161-1.

Senguttuvan, S. et al. (2022) 'Polypyrrole / zeolite composite – a nanoadsorbent for reactive dyes removal from synthetic solution', *Chemosphere*, 287(P2), p. 132164. doi:10.1016/j.chemosphere.2021.132164.

Singh, A. et al. (2023) 'Dose-dependent cytotoxicity against lung cancer cells via green synthesized ZnFe2O4/cellulose nanocomposites', *E-Polymers*, 23(1), pp. 1–11. doi:10.1515/epoly-2023–0113.

Singh, A. and Tomar, R. (2024) 'Efficient removal of crystal violet dye from water using zinc ferrite - polyaniline nanocomposites', *Environmental Monitoring and Assessment*, pp. 1–18. doi:10.1007/s10661-024-12686-z.

Solomon, N.O. et al. (2024) 'Sustainable nanomaterials' role in green supply chains and environmental sustainability', *Engineering Science & Technology Journal*, 5(5), pp. 1678–1694. doi:10.51594/estj/v5i5.1136.

Sutar, R.S. et al. (2020) 'Superhydrophobic PVC/SiO2 coating for self-cleaning application', *Macromolecular Symposia*, 393(1). doi:10.1002/masy.202000034.

Tajik, S. et al. (2022) 'Recent developments in polymer nanocomposite-based electrochemical sensors for detecting environmental pollutants', *Industrial & Engineering Chemistry Research*. doi:10.1021/acs.iecr.0c04952.

Tariq, A. et al. (2023) 'Grafting the ferrites of cobalt and zinc on MWCNTs for adsorption of crystal violet', *International Journal of Environmental Science and Technology*, 20(11), pp. 12465–12480. doi:10.1007/s13762-023-04857-1.

Thomas, B. et al. (2019) 'Nanocellulose, a versatile green platform: from biosources to materials and their applications', *Chemical Reviews*, 118(24), pp.11575–11625.

Torres-Giner, S. (2020) *Environmentally Friendly Polymers and Polymer Composites*. MDPI, pp. 1–6.

Torvi, A., Naik, S. and Kariduraganavar, M. (2018) 'Development of supercapacitor systems based on binary and ternary nanocomposites using chitosan, graphene and polyaniline', Chemical Data Collections [Preprint]. doi:10.1016/j.cdc.2018.11.003.

Toto, E., Laurenzi, S. and Santonicola, M.G. (2022) 'Recent trends in graphene / polymer nanocomposites for sensing devices: synthesis and applications in environmental and human health monitoring', *Polymers*, 14(5), p.1030.

Vafaeva, K.M. et al. (2024) 'Polymer matrix nanocomposites for sustainable packaging: a green approach', in E3S Web of Conferences. EDP Sciences, vol. 537, p. 08001.

Vattikuti, S.V.P. et al. (2024) 'Enhanced photocatalytic and electrochemical performance of MOF-derived NiO-ZnO oxide composites for wastewater treatment and sustainable energy storage', International Journal of Energy Research, 2024. doi:10.1155/2024/4589047.

Vignesh, M. and Moorthi, P.V. (2017) 'An overview of naturally synthesized metallic nanoparticles', *Journal of Applied Pharmaceutical Science*, 7(6), pp. 229–237. doi:10.7324/JAPS.2017.70634.

Wróblewska-Krepsztul, J. et al. (2018) 'Recent progress in biodegradable polymers and nanocomposite- based packaging materials for sustainable environment', International Journal of Polymer Analysis and Characterization, (February 2019). doi:10.1080/1023666X.2018.1455382.

Zenasni, M., Kiari, M. and Bakkour, Y. (2023) 'Synthesis, characterization, and enhanced electrochemical behavior of polypyrrole doped ZrO 2 – ZnO electrode materials for supercapacitor applications', Frontiers in Energy Research, 2023(August), pp. 1–11. doi:10.3389/fenrg.2023.1244699.

Zhao, X. et al. (2011) 'Polymer-supported nanocomposites for environmental application: a review', *Chemical Engineering Journal*, 170, pp. 381–394. doi:10.1016/j.cej.2011.02.071.

9 Economic and Commercial Status of Polymer Nanocomposites

Mohammad Harun-Ur-Rashid, Abu Bin Imran, and Md. Abu Bin Hasan Susan

9.1 INTRODUCTION

Polymer nanocomposites (PNCs) are a ground-breaking innovation in materials science, providing improved characteristics and flexible applications across a wide range of fields. These novel materials are made up of a polymer matrix included with nanoparticles, which greatly enhance the mechanical, thermal, and electrical characteristics of the underlying polymer. The notion of combining nanotechnology with polymers first appeared in the late twentieth century, and it has evolved rapidly owing to its transformational potential in the development of high-performing materials. The continuous advancements in nanocomposite technology have propelled its significance in modern industries, making it a critical area of research and development (R&D). PNCs have revolutionized multiple sectors, including automotive (Harun-Ur-Rashid & Imran, 2023b; Harun-Ur-Rashid, Imran, Susan, & Hasan, 2023), aerospace (Harun-Ur-Rashid & Imran, 2023a), electronics (Harun-Ur-Rashid, Foyez, & Imran, 2021; Harun-Ur-Rashid & Imran, 2022; M. J. Islam, Pal, Abdullah, & Harun-Ur-Rashid, 2024; M. J. Islam, Pal, Harun-Ur-Rashid, et al., 2024), consumer goods (M. Harun-Ur-Rashid, Imran, & Susan, 2023), environmental monitoring and safety (Harun-Ur-Rashid, Imran, & Susan, 2024a, 2024b), smart materials and sensing (Harun-Ur-Rashid, Jahan, et al., 2024; Harun-Ur-Rashid, Naidu Krishna, Golla, & Bin Imran, 2024; Imran, Harun-Ur-Rashid, & Takeoka, 2019), catalysis (Harun-Ur-Rashid, Pal, & Imran, 2024; Hassan et al., 2024), food packaging (Harun-Ur-Rashid & Imran, 2019; Harun-Ur-Rashid, Imran, & Susan, 2023), and healthcare (Harun-Ur-Rashid, Foyez, Jahan, Pal, & Imran, 2022; Harun-Ur-Rashid, Jahan, Foyez, & Imran, 2023; Harun-Ur-Rashid, Jahan, & Imran, 2023). In the automotive industry, these materials contribute to developing lightweight, durable components that enhance fuel efficiency and reduce emissions. The aerospace sector benefits from their exceptional strength-to-weight ratio, which is crucial for improving aircraft performance and safety. PNCs play a vital role in miniaturization and improving device functionality in electronics. The consumer goods and packaging industries leverage their durability and barrier properties to create advanced packaging solutions. Furthermore, the medical field utilizes PNCs for drug delivery systems, diagnostic devices, and biomedical implants, underscoring their versatility and impact on modern technology.

DOI: 10.1201/9781003527602-9

Despite significant advancements and applications of PNCs, the full realization of their potential is hindered by certain constraints. The scalability of synthesis and fabrication techniques often face challenges in maintaining uniform dispersion of nanoparticles and achieving consistent material properties at a large scale. In addition, there is a need for more comprehensive investigations on long-term impact of nanocomposites on the environment and human health, particularly regarding their degradation and potential toxicity. The development of cost-effective production methods that can make PNCs more accessible for widespread commercial use is also critically required. This chapter aims at comprehensively analyzing the economic and commercial status of PNCs for advancing the field and ensuring the sustainable and safe use of PNCs. It explores the fundamentals of nanotechnology and nanomaterials and investigates the properties and characteristics that make PNCs superior to traditional composites. Market analysis, highlighting the global market trends, key players, and competitive landscape, has been made. Furthermore, it examines the economic impact and diverse applications of PNCs across various industries, offering insights into their commercial potential and future prospects. The ultimate goal is to provide with strategic recommendations for stakeholders and industry participants, emphasizing the importance of innovation, sustainability, and regulatory compliance in driving the growth of this cutting-edge field and final thoughts on the future of PNCs in the global economy.

9.2 FUNDAMENTALS OF NANOTECHNOLOGY AND NANOMATERIALS

9.2.1 Introduction to Nanotechnology

Nanotechnology is a multidisciplinary science that involves the manipulation of matter at the atomic and molecular scales, typically below 100 nm. It encompasses the design, production, and application of structures, devices, and systems by controlling shape and size at the nanoscale. This field has revolutionized various sectors, including medicine, electronics, energy, and materials science, due to the unique properties exhibited by materials at the nanoscale. The key concepts of nanotechnology include the quantum mechanical effects that dominate at the nanoscale, leading to properties significantly different from those observed in bulk materials. This is exploited to manipulate individual atoms and molecules to develop materials with controllable and superior properties for specific applications. Key principles include self-assembly, nanoscale characterization, and the integration of nanostructures into larger systems.

9.2.2 Types of Nanomaterials

9.2.2.1 Carbon-Based Nanomaterials

Carbon-based nanomaterials, such as graphene and carbon nanotubes, exhibit exceptional mechanical, electrical, and thermal properties. Graphene, a single layer of carbon atoms arranged in a hexagonal lattice, gives extraordinary electrical conductivity, mechanical strength, and flexibility. Carbon nanotubes, cylindrical carbon molecules embedded in graphene covers in a lattice formation, also demonstrate remarkable electrical conductivity and strength, making them suitable for various applications from electronics to materials reinforcement (Chinnamayan, Nazim, & Palanichamy, 2024).

9.2.2.2 Metal and Metal Oxide Nanoparticles

Metal nanoparticles, including gold, silver, and platinum, exhibit unique electrical, optical, and catalytic properties owing to their high surface area-to-volume ratio and quantum effects. Metal oxide nanoparticles, such as TiO_2 and ZnO, are widely used in applications ranging from photocatalysis to UV protection in sunscreens. The morphology of these nanoparticles can be precisely controlled to tailor their properties for desired applications (Shubhadarshinee et al., 2024).

9.2.2.3 Polymer-Based Nanomaterials

Polymer-based nanomaterials are polymer matrices embedded with nanoparticles, such as silica, clay, or metal oxides. These materials combine the advantageous properties of polymers, such as flexibility and processability, with the enhanced mechanical, thermal, and barrier properties imparted by the nanoparticles. They are extensively used in PNCs, offering superior performance in applications such as packaging, automotive parts, and biomedical devices.

9.2.3 Properties of Nanomaterials

Compared to their bulk counterparts, nanomaterials exhibit distinct mechanical properties, including increased strength, hardness, and elasticity. Nanomaterials such as graphene and carbon nanotubes show exceptional conductivity, making them ideal for electronic devices and conductive composites. They often display enhanced heat resistance and conductivity, which are beneficial for thermal management applications. Nanoparticles can exhibit unique optical behaviors, such as surface plasmon resonance in metal nanoparticles, which is exploited in applications like sensing and imaging. Functionalization through modification of the surface of nanoparticles with specific chemical groups to enhance their compatibility with different matrices or to impart new functionalities is particularly important in applications such as drug delivery, where surface modifications can improve biocompatibility and targeting ability. The ability to precisely control surface chemistry allows for the customization of nanomaterials for specific applications, enhancing their utility and effectiveness. Table 9.1 compares various key nanomaterials, detailing their composition, key properties, and diverse applications.

9.3 FUNDAMENTALS OF POLYMER NANOCOMPOSITES

9.3.1 Overview of PNCs

The primary components of PNCs include the polymer matrix, nanofillers, and sometimes additional additives to improve compatibility and processing. The unique characteristics imparted by nanomaterials to the polymer matrices enable PNCs to

TABLE 9.1
Overview of Key Nanomaterials and Their Properties

Nanomaterial	Composition	Key Properties	Applications
Graphene	Carbon atoms in a hexagonal lattice	High electrical conductivity, mechanical strength, flexibility, large surface area	Electronics, composites, sensors, energy storage, medical devices, coatings, membranes
Carbon nanotubes	Cylindrical carbon structures	Exceptional strength, electrical conductivity, thermal stability, high aspect ratio	Reinforcements, electronics, conductive films, nanomedicine, energy storage, sensors
Gold nanoparticles	Gold atoms	Optical properties (surface plasmon resonance), biocompatibility, chemical stability	Medical imaging, drug delivery, catalysis, biosensors, photothermal therapy
Silver nanoparticles	Silver atoms	Antibacterial properties, electrical conductivity, optical properties	Antimicrobial coatings, electronics, textiles, medical devices, water treatment
Titanium dioxide nanoparticles	Titanium and oxygen atoms	Photocatalytic activity, UV absorption, chemical stability, non-toxic	Sunscreens, water treatment, self-cleaning surfaces, photocatalysis, pigments
Silica nanoparticles	Silicon and oxygen atoms	High surface area, thermal stability, low toxicity, biocompatibility	Reinforcements, drug delivery, catalysis, biomedical applications, coatings
Zinc oxide nanoparticles	Zinc and oxygen atoms	UV absorption, antibacterial properties, semiconductor properties	Sunscreens, medical applications, rubber industry, photocatalysis, sensors
Quantum dots	Semiconductor materials (e.g., CdSe, PbS)	Unique optical properties (size-dependent fluorescence), high quantum yield	Medical imaging, display technologies, photovoltaics, biosensors
Magnetic nanoparticles	Iron oxide, cobalt, nickel	Magnetic properties, biocompatibility, ease of functionalization	Magnetic resonance imaging (MRI), targeted drug delivery, data storage, environmental remediation
Fullerenes	Carbon atoms in a spherical shape (C60)	High electron affinity, antioxidant properties, versatility in functionalization	Drug delivery, electronic devices, photovoltaics, antioxidants
Nanoclays	Layered silicates	High aspect ratio, barrier properties, reinforcement capability	PNCs, coatings, flame retardants, rheology modifiers
Nanocellulose	Cellulose fibers	Biodegradability, high mechanical strength, large surface area	Packaging, biomedical applications, composites, rheology modifiers
Ceramic nanoparticles	Metal oxides, carbides, nitrides	High hardness, thermal stability, chemical resistance	Catalysis, coatings, structural materials, biomedical applications

perform functions that traditional materials cannot achieve. PNCs are widely used to improve the mechanical properties of materials, such as strength, stiffness, and toughness. Incorporating nanoscale fillers such as silica particles can significantly enhance the performance of polymer matrices. For example, the use of vinyl-modified narrow-dispersed silica cross-linkers has been shown to improve the mechanical properties of acrylamide-based hydrogels (Md Rezaul Karim, Harun-Ur-Rashid, & Imran, 2023). The development of highly stretchable and flexible materials is another significant application of PNCs. Researchers have developed hydrogels that exhibit remarkable stretchability by varying the size of vinyl-modified narrow-dispersed silica particles as cross-linkers (Mohammad Rezaul Karim, Harun-Ur-Rashid, & Imran, 2020). PNCs can be engineered to exhibit structural colors, which arise from the interaction of light with periodic nanostructures rather than from pigments. These materials can be used in applications such as displays, sensors, and decorative coatings. Structurally colored secondary particles composed of black and white colloidal particles can create vibrant colors without using dyes (Takeoka et al., 2013). PNCs are also used to develop smart materials responding to external stimuli, such as temperature, pH, or light. Stimuli-responsive materials can change their properties or structure in response to environmental changes, making them suitable for applications in sensors, actuators, and drug delivery systems. Template synthesis methods have been employed to fabricate stimuli-responsive, angle-independent structurally colored smart materials (Harun-Ur-Rashid, Imran, et al., 2009). In the biomedical field, PNCs are used to create advanced materials for drug delivery, tissue engineering, and medical devices. High-density poly(N-isopropylacrylamide) (PNIPA) brushes on silicon surfaces, synthesized using atom transfer radical polymerization, exemplify the use of PNCs in biomedical applications (Huda et al., 2010). PNCs can be designed to exhibit tunable optical properties, which help create photonic crystals and optical devices. Structural colored gels, for instance, can be engineered to act as tunable soft photonic crystals, allowing for the control of light propagation and reflection (Harun-Ur-Rashid, Seki, & Takeoka, 2009).

9.3.2 Properties and Characteristics

PNCs exhibit significantly improved properties compared to their base polymers. Mechanically, they offer enhanced strength, stiffness, and toughness due to the reinforcement provided by the nanofillers. PNCs demonstrate superior thermal stability and conductivity, making them suitable for high-temperature applications. Certain PNCs exhibit increased conductivity, which is beneficial for applications in electronics and electrical devices. Compared to traditional composites, which use micron-sized fillers, PNCs show remarkable improvements due to the high surface area-to-volume ratio of the nanofillers. This high surface area allows for better interaction between the polymer matrix and the fillers, leading to more effective stress transfer and improved properties. Additionally, PNCs often require lower filler content to achieve similar or better enhancements, resulting in lighter materials with superior performance.

9.3.3 Synthesis and Fabrication Techniques

The synthesis and fabrication of PNCs involve several methods, each with its own advantages and challenges.

9.3.3.1 Solution Mixing

This involves dissolving the polymer and the nanofillers in a common solvent, followed by mixing and solvent evaporation. It is widely used for its simplicity and effectiveness in achieving uniform dispersion of nanofillers.

9.3.3.2 *In Situ* Polymerization

Nanofillers are dispersed in a monomer solution, and polymerization is initiated to form the polymer matrix around the fillers. This method ensures strong interaction between the polymer and the nanofillers, resulting in excellent properties.

9.3.3.3 Melt Mixing

Melt mixing involves blending the polymer and nanofillers in a molten state using high-shear mixing equipment such as extruders. This is advantageous for its scalability and compatibility with existing polymer processing techniques.

9.3.3.4 Sol–Gel Process

This involves the transition of a system from a liquid "sol" into a solid "gel" phase. It is beneficial for incorporating inorganic nanofillers into a polymer matrix.

9.3.3.5 Challenges and Solutions in Fabrication

Fabricating PNCs presents several challenges, including achieving uniform dispersion of nanofillers, maintaining the stability of the nanocomposite, and ensuring compatibility between the polymer and the fillers. Solutions to these challenges involve surface functionalization of nanofillers, optimizing processing conditions, and using compatibilizers to enhance interfacial interactions. Table 9.2 provides an overview of various synthesis techniques for PNCs, detailing their descriptions, advantages, disadvantages, and typical applications.

9.4 MARKET ANALYSIS OF POLYMER NANOCOMPOSITES

9.4.1 Global Market Overview

9.4.1.1 Market Size and Growth Trends

The global market of PNCs was valued at USD 8.9 billion in 2022 and is projected to grow to USD 44.2 billion by 2032, exhibiting a compound annual growth rate (CAGR) of 19.50% during the forecast period (2023–2032). The market growth is driven by rising applications in electronics, automotive, aerospace, and other end-user industries, coupled with advancements in nanotechnology and material science.

TABLE 9.2
The Comparative Overview of Common Synthesis Techniques for PNCs

Synthesis Technique	Description	Advantages	Disadvantages	Typical Applications	References
Solution mixing	Dissolving polymer and nanofillers in a common solvent, followed by mixing and solvent evaporation	Simple, effective for uniform dispersion, easy to implement	Solvent removal can be challenging, limited to soluble polymers	Coatings, films, biomedical applications	Kuila et al. (2011)
In Situ polymerization	Dispersing nanofillers in monomer solution and initiating polymerization	Strong interaction between polymer and nanofillers, excellent properties	Complex process control, potential for incomplete polymerization	High-performance composites, structural materials	Mao and Wang (2020)
Melt mixing	Blending polymer and nanofillers in molten state using high-shear mixing equipment	Scalable, compatible with existing polymer processing techniques, no solvents required	High processing temperatures can degrade heat-sensitive nanofillers	Automotive parts, packaging materials	J. Banerjee and Dutta (2019)
Sol-gel process	Sol–gel transition	Useful for incorporating inorganic nanofillers, can form hybrid materials	Requires precise control of process conditions, potential for brittleness in final product	Optical coatings, ceramic-polymer composites	Amiri and Rahimi (2016)
Electrospinning	Using an electric field to draw polymer solutions into fine fibers	Produces nanofibers with high surface area-to-volume ratio, controllable fiber diameter	Limited to polymers that can be dissolved or melted, high voltage required	Filtration, tissue engineering scaffolds	Leonés, Lieblich, Benavente, Gonzalez, and Peponi (2020)

(*Continued*)

TABLE 9.2 (*Continued*)
The Comparative Overview of Common Synthesis Techniques for PNCs

Synthesis Technique	Description	Advantages	Disadvantages	Typical Applications	References
Layer-by-layer assembly	Alternately depositing layers of polymers and nanofillers	Precise control over composition and thickness, versatile for various materials	Time-consuming, may require multiple deposition cycles	Thin films, sensors, drug delivery systems	Soler (2018)
Intercalation and exfoliation	Inserting polymer chains between layers of nanofillers (e.g., clay) and separating them	Improved dispersion of nanofillers, enhanced mechanical properties	May require chemical modifications, potential for incomplete exfoliation	Barrier materials, flame-retardant coatings	Bousmina (2006)
High-pressure homogenization	Applying high pressure to mix polymer and nanofillers uniformly	Effective for creating homogeneous dispersions, suitable for large-scale production	Requires specialized equipment, potential for nanofiller agglomeration	Food packaging, cosmetic formulations	Wu et al. (2022)
Ball milling	Mechanical process involving grinding polymer and nanofillers together	Simple, scalable, can be used for a wide range of polymers and fillers	Potential for contamination, not suitable for heat-sensitive materials	Polymer composites, battery electrodes	Bozeya, Makableh, Al-Mezead, and Abu-Zurayk (2024)
Chemical vapor deposition (CVD)	Depositing nanofillers onto polymer substrates from a vapor phase	Produces high-purity nanocomposites, controllable nanofiller distribution	Requires high temperatures, complex equipment	Electronic devices, protective coatings	Mukherjee et al. (2023)

9.4.1.2 Key Regions and Market Segmentation

Key regions for the market include Asia-Pacific, Europe, North America, and the rest of the world. North America, particularly the United States, held the largest market share in 2022, driven by substantial investments in R&D and the presence of key industry players. The Asia-Pacific region is expected to register the fastest growth due to rapid industrialization and high demand from the automotive and industrial sectors in China and India.

9.4.2 Key Players and Competitive Landscape

9.4.2.1 Major Companies and Manufacturers

Several key players dominate the global market, including Arkema Group (known for its innovative materials and specialty chemicals, Arkema is a leading manufacturer of the PNCs), Evonik Industries AG (a major player with a strong focus on developing high-performance polymer solutions), Nanocyl SA (specializes in carbon nanotube-based composites), Nanophase Technologies Corporation (focuses on advanced materials and nanotechnology), RTP Company (offers a wide range of custom-engineered thermoplastics), Showa Denko KK (produces a variety of nanocomposite materials), Hybrid Plastics Inc. (known for its polyhedral oligomeric silsesquioxanes (POSS) nanocomposites), Inframat Corporation, Powdermet Inc., and Nanocor Inc. These companies are leading the market by continuously investing in R&D for expansion of the product lines and enhancement of technological capabilities.

9.4.2.2 Market Shares and Competitive Strategies

The competitive landscape is characterized by strategic initiatives such as mergers and acquisitions, partnerships, and product innovations. For example, the acquisition of Agiplast of Arkema has strengthened its position in high-performance polymer solutions. Innovations such as Novolen Pure polypropylene technology of Lummus Novolen Technology GmbH highlight ongoing efforts to enhance material performance and expand applications.

9.4.3 Market Drivers and Restraints

9.4.3.1 Technological Advancements

Technological advancements in nanotechnology and material science are crucial drivers of the market of PNCs. Innovations like self-regulating heating cables by nVent Electric PLC and other cutting-edge developments are propelling market growth by enhancing product performance and expanding application areas.

9.4.3.2 Economic Factors

Economic growth, increased industrialization, and rising disposable incomes contribute to expanding the market. The cost-effectiveness and long-term benefits of using PNCs in various applications drive their adoption across industries.

TABLE 9.3
Leading Companies in the Market of PNCs

Company	Key Products and Technologies	Market Share	Strategic Initiatives and Innovations
Arkema group	Nanostrength®, Graphistrength®	20%	Acquisition of Agiplast, development of sustainable nanocomposites
Evonik industries AG	High-performance polymers for 3D printing	15%	Collaboration with Lehvoss, focus on fiber-reinforced materials
Nanocyl SA	Carbon nanotube-based composites	10%	Product innovation, expansion in Asia-Pacific
Nanophase technologies	Advanced nanomaterials	8%	R&D in high-performance applications
RTP company	Custom-engineered thermoplastics	7%	Investment in new product development and market expansion
Showa denko KK	Various nanocomposite materials	5%	Development of cutting-edge nanocomposite solutions
Hybrid plastics Inc.	POSS nanocomposites	5%	Focus on enhancing material properties through innovative formulations
Others	Various regional and niche players	30%	Diverse product portfolios, focus on local markets

9.4.3.3 Regulatory Environment

Stringent regulations regarding the safety and environmental impact of nanomaterials influence the market. Compliance with these regulations necessitates continuous innovation and adaptation by manufacturers, ensuring the development of safe and sustainable nanocomposite technologies.

Table 9.3 provides an overview of the market of PNCs, highlighting the key products, market shares, and strategic initiatives of the major companies. Arkema Group, Evonik Industries AG, and Nanocyl SA lead the market with innovations in high-performance polymers and carbon nanotube-based composites. Strategic initiatives such as mergers, acquisitions, and product development are driving their market positions. The table also includes other major companies, focusing on R&D to expand their product lines and enhance technological capabilities.

9.5 ECONOMIC IMPACT AND APPLICATIONS

9.5.1 Automotive Industry

PNCs offer substantial benefits in the automotive industry, primarily through reduction of weight, enhancement of mechanical properties, and improvement of fuel efficiency (Chandrasekaran, Rameshbabu, & Prakash, 2024; Orji, Patricia, Sunday, & Olawale, 2024; Vinodhini & Xavier, 2024). By incorporating nanomaterials such as carbon nanotubes and nanoclays, PNCs provide superior strength and durability compared to traditional composites. This results in lighter vehicle components, such as bumpers, dashboards, and structural parts, contributing to overall vehicle performance and reduced emissions. Numerous case studies highlight the successful integration of PNCs in automotive applications. For example, Ford Motor Company has utilized nanocomposites in their F-150 trucks to enhance fuel efficiency and durability (Mitaľová, Mitaľ, & Simkulet, 2023). The use of PNCs in automotive parts such as engine covers and fuel line components has demonstrated significant improvements in performance and longevity, resulting in increased market adoption and customer satisfaction source.

9.5.2 Aerospace and Defense

In the aerospace and defense sectors, PNCs contribute to significant enhancement of performance and cost savings (Bag et al., 2024; Ranjbar, Modaberi, & Ashhari, 2024). The lightweight nature of PNCs, combined with their high strength-to-weight ratio, allows for developing more efficient aircraft and defense systems. These materials also provide improved thermal and barrier properties, which are essential for aerospace applications. PNCs are used in various aerospace and defense applications, including aircraft interiors, structural components, and protective coatings. For instance, Boeing has incorporated nanocomposites in constructing aircraft wings and fuselage panels (Joshi & Chatterjee, 2016), enhancing fuel efficiency and reducing maintenance costs. Similarly, in defense, PNCs are employed in developing lightweight armor and advanced sensor systems source.

9.5.3 Electronics and Electrical Industries

The role of PNCs in the electronics and electrical industries is crucial since they enable miniaturization of components and improve overall performance (Tamjid et al., 2024; Yang et al., 2024). Incorporating graphene and carbon nanotubes enhances the electrical conductivity and thermal management of electronic devices, leading to more efficient and compact product sources. PNCs are in high demand in electronics since devices need to be faster, smaller, and more efficient. Future trends indicate a growing adoption of PNCs in flexible electronics, wearable devices, and advanced computing technologies. Companies such as Samsung and Intel are making significant investments in R&D to integrate PNCs into next-generation electronic product sources.

9.5.4 Consumer Goods and Packaging

PNCs offer significant advantages in the consumer goods (Choudhury, Bindra, Singh, Singh, & Nayak, 2022) and packaging industries (Banerjee, Kapoor, Awasthi, & Rajasulochana, 2024) by enhancing the durability and functionality of products. Nanocomposite materials provide superior barrier properties against moisture, gases, and UV radiation, extending the shelf life of packaged goods and improving the performance of consumer product sources. The market for PNCs in consumer goods and packaging is growing due to increasing consumer demand for materials exhibiting high performance and sustainability. Innovations in nanocomposite packaging, such as biodegradable and recyclable materials, are gaining traction in the market.

Companies such as Unilever and Nestlé are exploring using PNCs to meet consumer preferences for eco-friendly packaging solutions source (Luo, Wu, & Zhi, 2016).

9.5.5 Medical and Healthcare

In the medical and healthcare fields, PNCs are utilized in various applications, including drug delivery systems (Mohammadpour-Haratbar, Zare, & Rhee, 2024), medical imaging (Andoh et al., 2024; Idumah, 2024), and diagnostics (Vydiam, Ahmad, & Mukherjee, 2024). The biocompatibility and functionalization capabilities of nanocomposites allow for targeted drug delivery, enhanced imaging contrast, and improved diagnostic accuracy source. The impact of PNCs in healthcare is significant, with ongoing research focusing on developing advanced nanocomposite-based medical devices and therapies. The prospects for PNCs in healthcare are promising, with potential applications in personalized medicine, regenerative therapies, and advanced diagnostic tools. Table 9.4 provides a detailed comparison of the applications of PNCs across various industries. It highlights the key applications, benefits, market potential, and comparative advantages over traditional materials, offering a comprehensive overview of the impact and potential of PNCs.

9.6 COMMERCIALIZATION STRATEGIES

9.6.1 Research and Development

The commercialization of PNCs is heavily driven by continuous R&D efforts. Emerging technologies and innovation trends are essential in enhancing the properties and applications of PNCs. Recent advancements include the development of multifunctional nanocomposites that combine mechanical strength with electrical conductivity, thermal stability, and biodegradability. Techniques such as electrospinning, 3D printing, and *in situ* polymerization are leading the way in producing advanced PNCs with tailored properties (Idumah, Hassan, & Ihuoma, 2019). Successful commercialization of PNCs requires robust collaboration between academia and industry. Academic institutions play a crucial role in fundamental research and early-stage technology development, while industrial partners bring expertise in scaling up production and market implementation. Initiatives such as joint research projects, industry-sponsored PhD programs, and technology transfer offices facilitate the seamless transition of innovative PNC technologies from the lab to the market (Islam & Ozcan, 2017).

9.6.2 Manufacturing and Production

Manufacturing PNCs at an industrial scale involves advanced production techniques to ensure that nanoparticles are evenly dispersed throughout the polymer matrix. Techniques such as melt blending, solution casting, and high-shear mixing are commonly used. Innovations in production, like continuous flow processing and automated manufacturing systems, have significantly improved the efficiency and consistency of PNC production (Kamal, Ashmawy, Shanmugam, Algazzar, &

TABLE 9.4
Applications of PNCs Across Different Industries

Industry	Key Applications	Benefits	Market Potential	Comparative Advantages Over Traditional Materials
Automotive	Lightweight components, structural parts, engine covers	Reduced weight, improved fuel efficiency, durability	High, driven by demand for efficient and eco-friendly vehicles	Lighter weight, better fuel efficiency, enhanced durability
Aerospace and defense	Aircraft interiors, structural components, protective coatings	Enhanced performance, cost savings, thermal management	High, due to increasing aerospace innovations and defense requirements	Higher strength-to-weight ratio, improved thermal stability, reduced maintenance costs
Electronics and electrical	Miniaturized components, flexible electronics, thermal management	Improved conductivity, compactness, performance	High, with growing demand for advanced electronic devices	Better electrical conductivity, improved thermal management, enabling miniaturization
Consumer goods and packaging	Packaging materials, durable consumer products	Superior barrier properties, sustainability	High, driven by consumer preferences for sustainable solutions	Improved barrier properties against moisture and gases, longer shelf life, eco-friendliness
Medical and healthcare	Drug delivery systems, medical imaging, diagnostics	Targeted delivery, enhanced imaging, accurate diagnostics	High, with ongoing advancements in medical technologies	Better biocompatibility, targeted drug delivery, enhanced imaging contrast
Energy	Batteries, fuel cells, solar panels	Increased efficiency, longer lifespan	High, due to growing demand for renewable energy solutions	Higher efficiency, improved storage capacity, enhanced durability
Construction	Reinforced concrete, coatings, insulation materials	Improved strength, durability, thermal insulation	Moderate to high, driven by infrastructure development	Enhanced mechanical properties, better thermal insulation, increased lifespan
Textiles	Smart fabrics, protective clothing, conductive fibers	Added functionality, improved comfort and safety	Moderate, driven by innovation in wearable technology	Enhanced strength, added functionalities like conductivity, improved comfort and safety
Sports and leisure	Lightweight equipment, performance gear, protective gear	Improved performance, enhanced durability	Moderate, driven by consumer demand for high-performance products	Lighter weight, better performance, increased durability

Elsheikh, 2021). One of the critical challenges in commercializing PNCs is achieving cost-effective scalability. While nanomaterials and advanced production techniques can be expensive, economies of scale and process optimization can reduce costs. Detailed cost analysis, including raw material costs, production expenses, and potential market prices, is essential for developing viable commercialization strategies. Companies must balance high performance with cost efficiency to make PNCs competitive (Kamal et al., 2021).

9.6.3 Marketing and Sales

Entering the market with PNCs requires well-planned strategies that address both the opportunities and challenges of the industry. Companies may adopt various market entry strategies, such as strategic alliances, joint ventures, and partnerships with established players. Targeting niche markets with high-value applications can also provide a foothold before expanding to broader markets. Understanding specific needs of target industries and pain points is crucial for successful market entry (Idumah et al., 2019). Effective branding and positioning are vital for distinguishing PNC products in a competitive market. Companies should emphasize the unique benefits of PNCs, such as enhanced performance, sustainability, and innovation. Building a strong brand identity that resonates with customers and stakeholders can enhance market penetration. Marketing efforts should highlight the advanced properties and diverse applications of PNCs through case studies, white papers, and industry events.

9.6.4 Regulatory and Environmental Considerations

Regulatory compliance is an important aspect of commercializing PNCs (Franz, Bott, & Störmer, 2020). Companies must adhere to international standards and regulations that govern the use of nanomaterials and polymers. These include ensuring product safety, environmental impact assessments, and obtaining necessary certifications. Compliance with standards such as ISO, REACH, and FDA guidelines helps gain market acceptance and avoid legal challenges. The environmental impact and sustainability of PNCs are increasingly important considerations for manufacturers and consumers (Carroccio et al., 2022; Coltelli, Morganti, & Lazzeri, 2023). Sustainable production practices, such as using renewable nanomaterials and reducing energy consumption, can enhance the environmental profile of PNCs. Companies should also focus on developing biodegradable and recyclable PNCs to address concerns about plastic pollution and resource depletion. Showcasing its commitment to sustainability can enhance reputation of a brand and attract in environmentally conscious clients.

9.7 FUTURE PROSPECTS AND TRENDS

9.7.1 Emerging Markets and Opportunities

The global market for PNCs is poised for substantial growth, with emerging markets presenting significant opportunities. High-potential regions include Asia-Pacific, particularly China and India, due to their rapid industrialization, robust manufacturing sectors, and increasing investments in nanotechnology. North America and Europe are key markets driven by technological advancements and high demand for automotive, aerospace, and electronics. Furthermore, sectors such as renewable energy, biomedical applications, and smart textiles are expected to see increased adoption of PNCs, offering lucrative opportunities for market expansion (Nepal, Haines, & Vaia, 2024).

9.7.2 Technological Advancements

The future of PNCs is marked by innovations and technological advancements. Artificial intelligence (AI) and machine learning are at the forefront of emerging technologies. These are being integrated into PNC research to optimize material properties and production processes. Innovations like self-healing nanocomposites, which can repair themselves after damage, and multifunctional nanocomposites, which combine properties such as conductivity and strength, are on the horizon. Advancements in 3D printing and additive manufacturing are also set to revolutionize the production of PNCs, enabling complex designs and customized applications.

9.7.3 Sustainability and Circular Economy

PNCs are critical in promoting sustainability (Paraye & Sarviya, 2024) and advancing the circular economy (Gupta & Sharma, 2024). By enhancing the durability and lifespan of products, PNCs contribute to resource efficiency and waste reduction. Developing biodegradable and recyclable nanocomposites is a significant trend, addressing environmental concerns and aligning with global sustainability goals. Furthermore, PNCs can improve the efficiency of renewable energy technologies, such as wind turbines and solar panels, contributing in the shift to a low-carbon economy.

9.7.4 Policy and Economic Forecast

Global policies and economic conditions significantly influence the PNC market (Shanmugam et al., 2021). Regulations aimed at reducing carbon emissions and promoting sustainable materials are driving the adoption of PNCs. Economic factors such as rising investments in nanotechnology research, increasing consumer demand for high-performance materials, and government incentives for green technologies are expected to boost the market. Conversely, economic downturns or stringent regulatory barriers could pose challenges. The substantial growth of the PNC market is anticipated to be supported by favorable policies and economic conditions.

Table 9.5 provides a detailed analysis of upcoming trends and innovations in the market of PNCs. It highlights the integration of AI and machine learning for optimizing material properties, developing self-healing and multifunctional nanocomposites, and advancements in 3D printing for customized designs. It emphasizes the move toward biodegradable and recyclable PNCs, the enhancement of renewable energy applications, and the positive impact of supportive regulatory policies and economic growth in emerging markets. These trends collectively predict robust market growth and evolution, driven by technological innovations and sustainability efforts.

TABLE 9.5
Future Trends and Predictions in the Market of PNCs

Trend/Prediction	Description	Impact on Market Growth	References
Integration of AI and machine learning	AI and machine learning to optimize material properties and production	Enhanced efficiency, customized applications, rapid R&D	Champa-Bujaico, Díez-Pascual, Lomas Redondo, and Garcia-Diaz (2024)
Self-healing nanocomposites	Development of nanocomposites that can self-heal and repair after damage	Increased durability and lifespan of products	Jamil et al. (2024)
Multifunctional nanocomposites	Combining multiple properties such as conductivity and mechanical strength	Broader range of applications, improved product performance	Jamil et al. (2024)
Advancements in 3D printing	Use of additive manufacturing for complex and customized PNC designs	Cost-effective production, innovation in design	Dananjaya, Marimuthu, Yang, Grace, and Abeykoon (2024), and Jankowska et al. (2024)
Biodegradable and recyclable PNCs	Development of environmentally friendly nanocomposites	Reduced environmental impact, alignment with sustainability	Avella, Salse, Sessini, Mincheva, and Lo Re (2024) and Dhayal et al. (2024)
Enhanced renewable energy applications	Improved performance of solar panels, wind turbines using PNCs	Contribution to low-carbon economy, increased energy efficiency	Al Ojeery and Farea (2024) and Morsi et al. (2024)
Regulatory support for green technologies	Government policies promoting sustainable materials	Increased market adoption, favorable investment environment	Kausar et al. (2024)
Economic growth in emerging markets	Industrialization and technological advancements in Asia-Pacific and beyond	Expansion of market reach, higher demand for PNCs	Nepal et al. (2024)

9.8 CONCLUSION

This chapter has underscored the transformative potential of PNCs across various industries, revealing significant growth potential driven by technological advancements and increasing demand. The global PNC market, valued at USD 8.9 billion in 2022, is projected to reach USD 44.2 billion by 2032. PNCs offer superior mechanical, thermal, and electrical properties to traditional composites, making them invaluable in automotive, aerospace, electronics, consumer goods, and healthcare applications. Effective commercialization strategies involve continuous investment in R&D, fostering collaborations between academia and industry, optimizing production processes, and prioritizing sustainable solutions. Future prospects are promising, with innovations in AI, self-healing materials, and biodegradable nanocomposites paving market growth. Stakeholders are encouraged to adopt targeted marketing strategies, comply with international regulations, and engage proactively with regulatory bodies to capitalize on the expanding opportunities in the PNC market. The future of PNCs in the global economy looks promising, driven by continuous innovations, expanding applications, and supportive policies, positioning PNCs as a cornerstone of modern material science and industry.

ACKNOWLEDGMENTS

MH thanks colleagues and collaborators of the Department of Chemistry, IUBAT, who provided critical feedback and support throughout the preparation of this manuscript. ABI and MABHS acknowledge the Ministry of Science and Technology for research funding on related projects.

REFERENCES

Al Ojeery, A., & Farea, M. O. (2024). Optimizing and boosting the physicochemical properties of some polymer nanocomposites for high-performance and flexible energy storage systems. *Journal of Energy Storage, 80*, 110309. doi:10.1016/j.est.2023.110309

Amiri, S., & Rahimi, A. (2016). Hybrid nanocomposite coating by sol–gel method: A review. *Iranian Polymer Journal, 25*(6), 559–577. doi:10.1007/s13726-016-0440-x

Andoh, V., Ocansey, D. K. W., Naveed, H., Wang, N., Chen, L., Chen, K., & Mao, F. (2024). The advancing role of nanocomposites in cancer diagnosis and treatment. *International Journal of Nanomedicine, 19*, 6099–6126. doi:10.2147/IJN.S471360

Avella, A., Salse, M., Sessini, V., Mincheva, R., & Lo Re, G. (2024). Reusable, recyclable, and biodegradable heat-shrinkable melt cross-linked poly(butylene adipate-co-terephthalate)/pulp biocomposites for polyvinyl chloride replacement. *ACS Sustainable Chemistry & Engineering, 12*(13), 5251–5262. doi:10.1021/acssuschemeng.4c00012

Bag, D. S., Kumar, A., Bansiwal, J. K., Lal, G., Tripathi, D. N., & Namburi, E. P. (2024). Polymer matrix composites (PMCs) for defence applications. In E. P. Namburi, R. J. H. Wanhill, & D. K. Setua (Eds.), *Novel Defence Functional and Engineering Materials (NDFEM) Volume 2: Engineering Materials for Defence Applications* (pp. 1–37). Springer Nature Singapore.

Banerjee, J., & Dutta, K. (2019). Melt-mixed carbon nanotubes/polymer nanocomposites. *Polymer Composites, 40*(12), 4473–4488. doi:10.1002/pc.25334

Banerjee, S., Kapoor, A., Awasthi, A., & Rajasulochana, P. (2024). Polymer nanocomposites for food packaging applications. In S. Mallakpour & C. M. Hussain (Eds.), *Handbook of Nanofillers* (pp. 1–25). Springer Nature Singapore.

Bousmina, M. (2006). Study of intercalation and exfoliation processes in polymer nanocomposites. *Macromolecules, 39*(12), 4259–4263. doi:10.1021/ma052647f

Bozeya, A., Makableh, Y. F., Al-Mezead, L. A., & Abu-Zurayk, R. (2024). Wet ball milling and hot press for the preparation of UHMWPE/modified MWCNTs nanocomposite with enhanced mechanical and thermal properties. *Polymer Bulletin, 81*(2), 1707–1727. doi:10.1007/s00289-023-04790-w

Carroccio, S. C., Scarfato, P., Bruno, E., Aprea, P., Dintcheva, N. T., & Filippone, G. (2022). Impact of nanoparticles on the environmental sustainability of polymer nanocomposites based on bioplastics or recycled plastics – A review of life-cycle assessment studies. *Journal of Cleaner Production, 335*, 130322. doi:10.1016/j.jclepro.2021.130322

Champa-Bujaico, E., Díez-Pascual, A. M., Lomas Redondo, A., & Garcia-Diaz, P. (2024). Optimization of mechanical properties of multiscale hybrid polymer nanocomposites: A combination of experimental and machine learning techniques. *Composites Part B: Engineering, 269*, 111099. doi:10.1016/j.compositesb.2023.111099

Chandrasekaran, P., Rameshbabu, V., & Prakash, C. (2024). Advancements in Basalt composite automobile bumpers and performance evaluation through finite element analysis. *Polymer Bulletin, 81*(7), 6073–6090. doi:10.1007/s00289-023-04985-1

Chinnamayan, S., Nazim, A., & Palanichamy, K. (2024). Carbon-based nanocomposites: A comprehensive review of their multifunctional applications. In V. Parvulescu & E. M. Anghel (Eds.), *Nanocomposites - Properties, Preparations and Applications* (pp. Ch. 35). IntechOpen.

Choudhury, M., Bindra, H. S., Singh, K., Singh, A. K., & Nayak, R. (2022). Antimicrobial polymeric composites in consumer goods and healthcare sector: A healthier way to prevent infection. *Polymers for Advanced Technologies, 33*(7), 1997–2024. doi:10.1002/pat.5660

Coltelli, M.-B., Morganti, P., & Lazzeri, A. (2023). Chapter 20- Sustainability assessment, environmental impact, and recycling strategies of biodegradable polymer nanocomposites. In K. Deshmukh & M. Pandey (Eds.), *Biodegradable and Biocompatible Polymer Nanocomposites* (pp. 699–737). Elsevier.

Dananjaya, V., Marimuthu, S., Yang, R., Grace, A. N., & Abeykoon, C. (2024). Synthesis, properties, applications, 3D printing and machine learning of graphene quantum dots in polymer nanocomposites. *Progress in Materials Science, 144*, 101282. doi:10.1016/j.pmatsci.2024.101282

Dhayal, V., Leel, N. S., Hasan, P. M. Z., Darwesh, R., Quraishi, A. M., Hashmi, S. Z., … Alvi, P. A. (2024). Recyclable CS/r-GO polymer nanocomposites: Green synthesis and characterizations. *Physica Status Solidi (A), 221*(9), 2300894. doi:10.1002/pssa.202300894

Franz, R., Bott, J., & Störmer, A. (2020). Considerations for and guidance to testing and evaluating migration/release of nanoparticles from polymer based nanocomposites. *Nanomaterials, 10*(6). doi:10.3390/nano10061113

Gupta, G., & Sharma, B. (2024). Chapter 16- Circular economy and upcoming horizons in the field of bionanocomposites. In B. Sharma, S. Thomas, P. K. Bajpai, K. Ghosal, & S. Shekhar (Eds.), *Advances in Bionanocomposites* (pp. 365–384). Elsevier.

Harun-Ur-Rashid, M., Foyez, T., & Imran, A. B. (2021). Fabrication of stretchable composite thin film for superconductor applications. In K. Pal (Ed.), *Sensors for Stretchable Electronics in Nanotechnology* (1st edn, pp. 63–78). CRC Press.

Harun-Ur-Rashid, M., Foyez, T., Jahan, I., Pal, K., & Imran, A. B. (2022). Rapid diagnosis of COVID-19 via nano-biosensor-implemented biomedical utilization: A systematic review. *RSC Advances, 12*(15), 9445–9465. doi:10.1039/D2RA01293F

Harun-Ur-Rashid, M., & Imran, A. B. (2019). Superabsorbent hydrogels from carboxymethyl cellulose. In M. I. H. Mondal (Ed.), *Carboxymethyl Cellulose* (Vol. 1, pp. 159). Nova Science Publisher, Inc.

Harun-Ur-Rashid, M., & Imran, A. B. (2022). Organometallic nanomaterials synthesis and sustainable green nanotechnology applications. In K. Pal (Ed.), *Green Nanoarchitectonics* (1st edn, pp. 249–270). Jenny Stanford Publishing.

Harun-Ur-Rashid, M., & Imran, A. B. (2023a). Engineered nanomaterials for energy conversion cells. In N. B. Singh, M. A. B. H. Susan, & R. G. Chaudhary (Eds.), *Applications of Emerging Nanomaterials and Nanotechnology* (Vol. Materials Research Foundations, pp. 103–126). Materials Research Forum LLC.

Harun-Ur-Rashid, M., & Imran, A. B. (2023b). Nanomaterials in the automobile sector. In N. B. Singh, M. A. B. H. Susan, & R. G. Chaudhary (Eds.), *Emerging Applications of Nanomaterials* (Vol. 141, pp. 124–150). Materials Research Foundations.

Harun-Ur-Rashid, M., Imran, A. B., Seki, T., Takeoka, Y., Ishii, M., & Nakamura, H. (2009). Template synthesis for stimuli-responsive angle independent structural colored smart materials. *Transactions of the Materials Research Society of Japan, 34*(2), 333–337. doi:10.14723/tmrsj.34.333

Harun-Ur-Rashid, M., Imran, A. B., & Susan, M. (2023). Prospective nanomaterials for food packaging and safety. In N. B. Singh, M. A. B. H. Susan, & R. G. Chaudhary (Eds.), *Emerging Applications of Nanomaterials, Materials Research Forum LLC* (Vol. 141, pp. 327–352). Materials Research Foundations.

Harun-Ur-Rashid, M., Imran, A. B., Susan, M., & Hasan, A. B. (2023). Green polymer nanocomposites in automotive and packaging industries. *Current Pharmaceutical Biotechnology, 24*(1), 145–163. doi:10.2174/1389201023666220506111027

Harun-Ur-Rashid, M., Imran, A. B., & Susan, M. A. B. H. (2023). Fire-resistant polymeric foams and their applications. In R. K. Gupta (Ed.), Polymeric Foams: Applications of Polymeric Foams (Vol. ACS Symposium Series, pp. pp 97–121). ACS Publications.

Harun-Ur-Rashid, M., Imran, A. B., & Susan, M. A. B. H. (2024a). Emerging 3D printed polymers and composites for water quality preservation. In *Reference Module in Materials Science and Materials* Engineering. Elsevier.

Harun-Ur-Rashid, M., Imran, A. B., & Susan, M. A. B. H. (2024b). Magnetic polymer nanocomposites for removal of dyes and metals from wastewater. In *Reference Module in Materials Science and Materials* Engineering. Elsevier.

Harun-Ur-Rashid, M., Jahan, I., Foyez, T., & Imran, A. B. (2023). Bio-inspired nanomaterials for micro/nanodevices: A new era in biomedical applications. *Micromachines, 14*(9). doi:10.3390/mi14091786

Harun-Ur-Rashid, M., Jahan, I., & Imran, A. B. (2023). Recent advancement of nanostructured materials for clinical challenges in vaccinology. In K. Pal (Ed.), *Nanovaccinology: Clinical Application of Nanostructured Materials Research to Translational Medicine* (pp. 135–160). Springer International Publishing.

Harun-Ur-Rashid, M., Jahan, I., Islam, M. J., Kumer, A., Huda, M. N., Imran, A. B., … Alsalhi, S. A. (2024). Global advances and smart innovations in supramolecular polymers. *Journal of Molecular Structure, 1304*, 137665. doi:10.1016/j.molstruc.2024.137665

Harun-Ur-Rashid, M., Naidu Krishna, S. B., Golla, N., & Bin Imran, A. (2024). Polyphenol oxidases: The future toward global sustainability. In P. Verma, K. Agrawal, & M. P. Shah (Eds.), *Function, Wastewater Remediation, and Biosensors* (pp. 309–344). De Gruyter.

Harun-Ur-Rashid, M., Pal, K., & Imran, A. B. (2024). Hybrid nanocomposite fabrication of nanocatalyst with enhanced and stable photocatalytic activity. *Topics in Catalysis, 67*, 17–45. doi:10.1007/s11244-023-01809-4

Harun-Ur-Rashid, M., Seki, T., & Takeoka, Y. (2009). Structural colored gels for tunable soft photonic crystals. *The Chemical Record, 9*(2), 87–105. doi:https://doi.org/10.1002/tcr.20169

Hassan, M. K., Karim, M. T., Biswas, P., Howlader, D., Harun-Ur-Rashid, M., & Kumer, A. (2024). Computational investigation for tetragonal crystals of Zn(GaS2)2, Zn(GaSe2)2, and Zn(GaTe2)2 photocatalysts for wastewater treatment: First principle approaches. *Advanced Journal of Chemistry-Section B: Natural Products and Medical Chemistry, 6*(1), 46–66. doi:10.48309/ajcb.2024.421434.1204

Huda, M. N., Seki, T., Suzuki, H., ANM, H. K., Harunur-Rashid, M., & Takeoka, Y. (2010). Characteristics of high-density poly (N-isopropylacrylamide)(PNIPA) brushes on silicon surface by atom transfer radical polymerization. *Transactions of the Materials Research Society of Japan, 35*(4), 845–848. doi:10.14723/tmrsj.35.845

Idumah, C. I. (2024). Phosphorene polymeric nanocomposites for biomedical applications: A review. *International Journal of Polymeric Materials and Polymeric Biomaterials, 73*(4), 292–309. doi:10.1080/00914037.2022.2158333

Idumah, C. I., Hassan, A., & Ihuoma, D. E. (2019). Recently emerging trends in polymer nanocomposites packaging materials. *Polymer-Plastics Technology and Materials, 58*(10), 1054–1109. doi:10.1080/03602559.2018.1542718

Imran, A. B., Harun-Ur-Rashid, M., & Takeoka, Y. (2019). Polyrotaxane actuators. In K. Asaka & H. Okuzaki (Eds.), *Soft Actuators: Materials, Modeling, Applications, and Future Perspectives* (2nd edn, pp. 81–147). Springer Nature.

Islam, M. J., Pal, K., Abdullah, M. M., & Harun-Ur-Rashid, M. (2024). Threshold bandgap of lead halide hybrid perovskites for making stable single crystal by NIR laser trapping. *Journal of Molecular Structure, 1301*, 137322. doi:10.1016/j.molstruc.2023.137322

Islam, M. J., Pal, K., Harun-Ur-Rashid, M., Kumar, A., Hossain, M. F., Asthana, N., … Malik, A. (2024). Optical maneuvering of photofunctioning hybrid perovskite for future photonics potential application. *Journal of Molecular Liquids, 408*, 125343. doi:10.1016/j.molliq.2024.125343

Islam, N., & Ozcan, S. (2017). Dynamics of Patent Collaboration: The Case of Nanocomposite Materials. Paper presented at the 2017 Portland International Conference on Management of Engineering and Technology (PICMET), Portland, United States, 9–13 July 2017.

Jamil, H., Faizan, M., Adeel, M., Jesionowski, T., Boczkaj, G., & Balčiūnaitė, A. (2024). Recent advances in polymer nanocomposites: Unveiling the frontier of shape memory and self-healing properties—A comprehensive review. *Molecules, 29*(6). doi:10.3390/molecules29061267

Jankowska, M., Lepcio, P., Chachaj-Brekiesz, A., Galek, M., Petko, F., Vozárik, A., & Ortyl, J. (2024). High-efficiency pentafluorostilbene-based photocatalysts dedicated to preparing fluorescent 3D printed polymer nanocomposites. *Virtual and Physical Prototyping, 19*(1), e2301030. doi:10.1080/17452759.2023.2301030

Joshi, M., & Chatterjee, U. (2016). 8- Polymer nanocomposite: An advanced material for aerospace applications. In S. Rana & R. Fangueiro (Eds.), *Advanced Composite Materials for Aerospace Engineering* (pp. 241–264). Woodhead Publishing.

Kamal, A., Ashmawy, M., Shanmugam, S., Algazzar, A. M., & Elsheikh, A. H. (2021). Fabrication techniques of polymeric nanocomposites: A comprehensive review. *Proceedings of the Institution of Mechanical Engineers, Part C: Journal of Mechanical Engineering Science, 236*(9), 4843–4861. doi:10.1177/09544062211055662

Karim, M. R., Harun-Ur-Rashid, M., & Imran, A. B. (2023). Effect of sizes of vinyl modified narrow-dispersed silica cross-linker on the mechanical properties of acrylamide based hydrogel. *Scientific Reports, 13*(1), 5089. doi:10.1038/s41598-023-32185-4

Karim, M. R., Harun-Ur-Rashid, M., & Imran, A. B. (2020). Highly stretchable hydrogel using vinyl modified narrow dispersed silica particles as cross-linker. *ChemistrySelect, 5*(34), 10556–10561. doi:10.1002/slct.202003044

Kausar, A., Ahmad, I., Aldaghri, O., Ibnaouf, K. H., Eisa, M. H., & Lam, T. D. (2024). Cutting-edge green nanoclay nanocomposites—Fundamentals and technological opportunities for packaging, dye removal, and biomedical sectors. *Nanocomposites, 10*(1), 172–196. doi:10.1080/20550324.2024.2336356

Kuila, T., Bose, S., Hong, C. E., Uddin, M. E., Khanra, P., Kim, N. H., & Lee, J. H. (2011). Preparation of functionalized graphene/linear low density polyethylene composites by a solution mixing method. *Carbon, 49*(3), 1033–1037. doi:https://doi.org/10.1016/j.carbon.2010.10.031

Leonés, A., Lieblich, M., Benavente, R., Gonzalez, J. L., & Peponi, L. (2020). Potential applications of magnesium-based polymeric nanocomposites obtained by electrospinning technique. *Nanomaterials, 10*(8). doi:10.3390/nano10081524

Luo, D., Wu, L., & Zhi, J. (2016). 10- Nanocomposite for food encapsulation packaging. In A. M. Grumezescu (Ed.), *Encapsulations* (pp. 421–454). Academic Press.

Mao, H.-n., & Wang, X.-g. (2020). Use of in-situ polymerization in the preparation of graphene / polymer nanocomposites. *New Carbon Materials, 35*(4), 336–343. doi:10.1016/S1872-5805(20)60493-0

Mitaľová, Z., Mitaľ, D., & Simkulet, V. (2023, 2023//). Application of Composite Materials Natural Fibers in Automotive Industry – Short Review. Paper presented at the EAI International Conference on Automation and Control in Theory and Practice, Cham.

Mohammadpour-Haratbar, A., Zare, Y., & Rhee, K. Y. (2024). A review on drug delivery systems containing polymer nanocomposites for breast cancer treatment. *Polymer Reviews, 64*(2), 490–527. doi:10.1080/15583724.2023.2262542

Morsi, M. A., Asnag, G. M., Assran, A. S., Alwafi, R., Tarabiah, A. E., Alshehri, N. A., … Saeed, A. (2024). Reinforced PEO/Cs polymers blend with Al2O3/TiO2 hybrid nanofillers: Nanocomposites for optoelectronics and energy storage. *Journal of Energy Storage, 88*, 111554. doi:10.1016/j.est.2024.111554

Mukherjee, A., Dianatdar, A., Gładysz, M. Z., Hemmatpour, H., Hendriksen, M., Rudolf, P., … Bose, R. K. (2023). Electrically conductive and highly stretchable piezoresistive polymer nanocomposites via oxidative chemical vapor deposition. *ACS Applied Materials & Interfaces, 15*(26), 31899–31916. doi:10.1021/acsami.3c06015

Nepal, D., Haines, J., & Vaia, R. A. (2024). Polymer nanocomposites: 35 years on. *MRS Bulletin, 49*(3), 236–246. doi:10.1557/s43577-023-00639-9

Orji, U. U., Patricia, P. A., Sunday, A. V., & Olawale, P. (2024). Development of polymer/carbon nanotubes incorporated sustainable materials for manufacturing of autobrake pad. *The International Journal of Advanced Manufacturing Technology, 132*(7), 3227–3236. doi:10.1007/s00170-024-13536-5

Paraye, P., & Sarviya, R. M. (2024). Advances in polymer composites, manufacturing, recycling, and sustainable practices. *Polymer-Plastics Technology and Materials, 63*(11), 1474–1497. doi:10.1080/25740881.2024.2339295

Ranjbar, Z., Modaberi, M. R., & Ashhari, S. (2024). 17- Polymer nanocomposite films and coatings in aerospace applications. In M. Pandey, K. Deshmukh, & C. M. Hussain (Eds.), *Polymer Nanocomposite Films and Coatings* (pp. 591–629). Woodhead Publishing.

Shanmugam, V., Mensah, R. A., Försth, M., Sas, G., Restás, Á., Addy, C., … Ramakrishna, S. (2021). Circular economy in biocomposite development: State-of-the-art, challenges and emerging trends. *Composites Part C: Open Access, 5*, 100138. doi:10.1016/j.jcomc.2021.100138

Shubhadarshinee, L., Mohapatra, P., Behera, S., Jali, B. R., Mohapatra, P., & Barick, A. K. (2024). Review on synthesis and characterization of metal nanoparticles doped carbon nanofillers based nanohybrids reinforced polyaniline nanocomposites. *Polymer-Plastics Technology and Materials, 63*(8), 1011–1035. doi:10.1080/25740881.2024.2314508

Soler, M. A. G. (2018). Layer-by-layer assembled iron oxide based polymeric nanocomposites. *Journal of Magnetism and Magnetic Materials, 467*, 37–48. doi:10.1016/j.jmmm.2018.07.035

Takeoka, Y., Yoshioka, S., Teshima, M., Takano, A., Harun-Ur-Rashid, M., & Seki, T. (2013). Structurally coloured secondary particles composed of black and white colloidal particles. *Scientific Reports, 3*(1), 2371. doi:10.1038/srep02371

Tamjid, E., Najafi, P., Khalili, M. A., Shokouhnejad, N., Karimi, M., & Sepahdoost, N. (2024). Review of sustainable, eco-friendly, and conductive polymer nanocomposites for electronic and thermal applications: Current status and future prospects. *Discover Nano, 19*(1), 29. doi:10.1186/s11671-024-03965-2

Vinodhini, S. P., & Xavier, J. R. (2024). Novel multifunctional nanocomposites containing graphitic carbon nitride/silanized Nb2O5 nanofillers for the protection of steel surfaces in the automobile industry. *Journal of Molecular Structure, 1306*, 137875. doi:10.1016/j.molstruc.2024.137875

Vydiam, K., Ahmad, J., & Mukherjee, S. (2024). Chapter 8- Development of metal–polymer composite nanomaterials for diagnosis and phototherapy. In M. Abbas, A. Atiq, M. Ovais, & M. R. Hamblin (Eds.), *Organic Nanomaterials for Cancer Phototheranostics* (pp. 171–194). Woodhead Publishing.

Wu, H., Zhou, W., Liu, Q., Cai, X., Qu, Z., Li, P., … Jia, X. (2022). High pressure homogenization of graphene and carbon nanotube for thermal conductive polyethylene composite with a low filler content. *Journal of Applied Polymer Science, 139*(12), 51838. doi:10.1002/app.51838

Yang, M., Guo, M., Xu, E., Ren, W., Wang, D., Li, S., … Shen, Y. (2024). Polymer nanocomposite dielectrics for capacitive energy storage. *Nature Nanotechnology, 19*(5), 588–603. doi:10.1038/s41565-023-01541-w

10 Tomorrow's World

How Polymer-Based Nanocomposites Lead the Nano Revolution

Israt Jahan Heme, Kawsar Ahmmed, Sumaiya Hasan Suma, and Md. Abu Bin Hasan Susan

10.1 INTRODUCTION

Nanotechnology is a fascinating area of research in the modern era and is rapidly advancing. It has a wide range of applications in areas such as physics, biology, chemistry, materials science, and even engineering and thus is often referred to as an interdisciplinary field. Nanocomposites (NCs) are materials with at least one dimension on the nanometer scale. Their size, dimension, shape, and geometry can determine their applications and design concepts (Yadav et al. 2020; Islam et al. 2020). They exhibit exceptional performance and offer unique design possibilities. These often consist of two or more components with varying physical and chemical properties, which are often separated by interfaces. This combination produces features that no single component could accomplish on its own. NCs are categorized into three classes based on the matrix or host material used: ceramic-based, metal-based, and polymer-based NCs (PNCs) (Thinh et al. 2020). PNCs, emerging as transformative, offer unprecedented capabilities that are poised to reshape industries and enhance everyday life. They integrate polymers with nanometer-scale fillers and combine the flexibility and processability of polymers with the superior properties of nanoparticles (NPs), resulting in materials that exhibit exceptional strength, durability, and functionality. The nano revolution, characterized by the manipulation of matter at the atomic and molecular scale, is fundamentally changing the landscape of material science. At the heart of this revolution are PNCs, which harness the unique properties of nanoscale fillers to overcome the limitations of traditional materials. The incorporation of NPs, such as carbon nanotubes (CNTs), graphene, and metal oxides, into polymer matrices enhances their properties, leading to innovations that extend across a multitude of applications. The potential of PNCs extends beyond their enhanced properties; they also offer solutions to pressing environmental issues. The development of biodegradable and environmentally friendly

DOI: 10.1201/9781003527602-10

NCs reflects a growing commitment to sustainable practices, aligning technological progress with ecological stewardship.

Traditional nanofillers have garnered significant attention in both academic and industrial circles. Depending on the matrix-nanofiller combination, PNCs can be tailored to specific needs. Recently, PNCs have sparked considerable research interest due to their ability to significantly improve various composite properties at minimal cost, leading to their integration into our daily lives. The compatibility between the polymer and nanofiller results in synergistic enhancements to the properties of composites, yielding characteristics that surpass those of the individual components. Figure 10.1 represents the classification of the key nanofillers and outline subcategories. The key characteristics of the nanofillers linked to their applications are highlighted by colored dashed lines, while shape and dimensional properties are depicted in a range of gray shades.

Carbon-based nanofillers like CNT and graphene are widely utilized in developing NCs because of their outstanding mechanical properties and electrical and thermal conductivities (Lau et al. 2006, Wang et al. 2009). CNT reinforces PNCs in the creation of bulletproof shields, jackets, shock absorbers, impact-resistant surfaces, and materials designed to withstand ballistic impacts (Coleman et al. 2006, Ma et al. 2010). The progress in advanced PNCs is leading to more sophisticated materials with tailored properties, opening up new possibilities for technology and industry leading to a sustainable future.

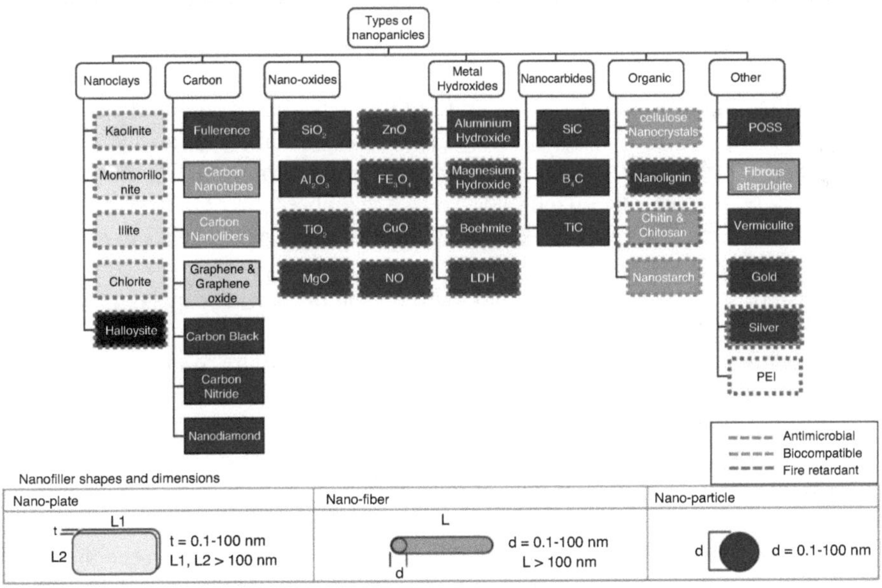

FIGURE 10.1 Categories of nanofillers and their respective subcategories (Doagou-Rad et al. 2020).

10.2 POLYMER-BASED NANOCOMPOSITES—CLASSIFICATION

NCs are composed of a matrix (metal, polymer, or ceramic) with reinforcing nanofillers such as NPs, nanofibers, or nanoplates (Singh et al. 2023). In PNCs, polymers are often reinforced with inorganic or organic fillers. The large surface area controls the characteristics of PNCs and regulates the contact level between the polymer and the filler. Due to their affordability, low cost, and sustainable usage, PNCs are becoming the most desirable and versatile materials worldwide. PNCs outperform conventional polymer composites by utilizing a small amount of nanofiller, which yields better results than using more microfillers. PNCs are classified based on the type of polymer matrix, nanofiller, and technique for synthesis. This section discusses different types of PNCs and their synthesis.

10.2.1 Polymer Matrix

The polymer itself is the main component of PNCs. Depending on the type of polymer matrix, PNCs can be classified as follows:

i. Thermoplastic: Thermoplastics are a popular choice for the matrix material in PNCs due to their processability and versatility. These polymers become soft and moldable when heated and solidify when cooled. Since there is no bond formation involved, their curing method is entirely recyclable. Common examples of thermoplastics include polycarbonate (PC), polyethylene (PE), polyvinyl chloride (PVC), polystyrene (PS), and polyamide (PA) (Cruz-Chú et al. 2020).
ii. Thermoset: In a curing process, polymer chains chemically cross-link to produce thermoset plastics, resulting in a three-dimensional network. This cross-linking gives the material rigidity and durability in an irreversible process. Thermosetting polymers such as polyurethanes (PU), polyesters, epoxy resin, and vulcanized materials (Stenzenberger 1993) serve as robust and stable matrices for PNCs, leveraging their unique properties to enhance the quality of the NCs.
iii. Elastomer: Elastomers, materials with elastic properties capable of extensive stretching without deforming are classified as viscoelastic due to their high viscosity. The addition of various nanofillers to elastomers significantly enhances their thermal, mechanical, fire resistance, barrier, and other properties. Thermoplastic elastomers, natural rubber, and synthetic rubber (e.g., silicone and neoprene) are examples of elastomer matrices (Maiti et al. 2008).
iv. Biodegradable Natural Polymer: While natural polymers may possess weak mechanical qualities, they have the potential for biological recognition, which can aid in cell attachment and activation. Introducing nanofillers into these polymers can improve their mechanical properties and other characteristics. Examples of biodegradable natural polymers include proteins (collagen, silk, and soy) and polysaccharides (starch, chitin/chitosan, alginate, and hyaluronic acid derivatives) (Armentano et al. 2010).

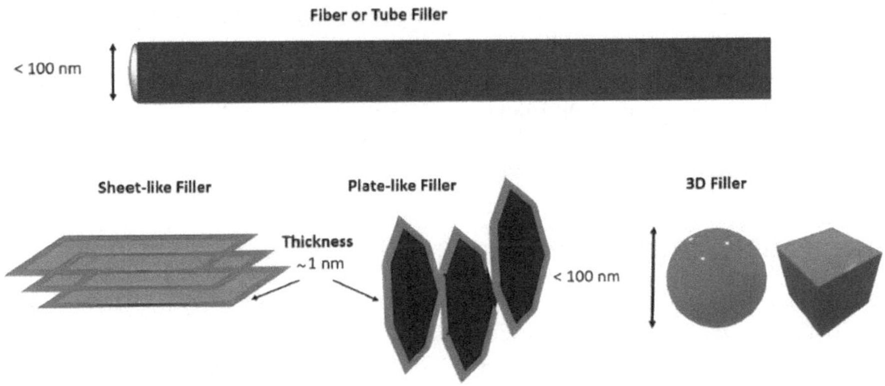

FIGURE 10.2 Different types of nanofillers for PNCs (Singh et al. 2023).

10.2.2 Dimension of Nanofillers

PNCs are often categorized on the basis of the nanofiller dimensions (Figure 10.2).

1. Nanofillers, with all dimensions below 100nm are defined as 0D nanofillers like carbon dots (CDs), quantum dots, and fullerenes, offer significant advantages in enhancing the properties of the PNCs.
2. Some nanofillers have only one dimension in the nanoscale range such as nanosheet of graphene or silicate and plate-like fillers. PNCs containing these nanofillers are often layered (Fu et al. 2019).
3. Nanofillers with two dimensions in the nanoscale range are also used in PNCs. They usually have an extended one-dimensional form due to the third dimension, which is larger than the former two dimensions. Examples include nanotubes and nanofibers.
4. PNCs may also contain three-dimensional (3D) nanofillers. 3D nanofillers have a more complex interconnected structure. This dimensionality imparts unique properties and functionalities to the PNCs. Nanoclays, zeolites, and nanogranules are examples of 3D nanofillers.

10.3 TECHNIQUES FOR SYNTHESIS

To improve the qualities of the final product, nanofillers are introduced into a polymer matrix during the manufacture of PNCs. Several techniques can be used for this purpose.

- In ex situ synthesis of the PNCs, the polymer and nanofillers are manufactured separately, and then nanofillers are added to the melted polymer or polymeric solution. Ex situ techniques ensure uniform dispersion in the polymer matrix and provide exact control over the characteristics of the nanofiller.

- *In situ* synthesis of PNCs involves incorporating nanofillers into the polymer matrix during the polymerization process.
- Precursors may be introduced into the voids of polymers, which might eventually transform into nanofillers.
- Precursors for nanofiller and monomers can be added at the same time in the system to fabricate the PNCs.

10.4 FABRICATION OF POLYMER-BASED NANOCOMPOSITES

The prerequisite for producing PNCs is the uniform dispersion of nanofillers throughout the matrix. Aggregation of nanofillers can occur, which affects the properties of the PNCs and limits their performance. Therefore, the fabrication of PNCs while retarding the aggregation of nanofillers is challenging. The following techniques are often used for fabricating PNCs (Tanahashi 2010).

10.4.1 Intercalation Method

Intercalation method involves the insertion or intercalation of nanofillers, such as clays or layered materials, between the layers of a polymer matrix. In order to make nanofillers more compatible with the polymer matrix, their surfaces are frequently modified. Chemical and mechanical methods are used in this regard.

Chemical methods involve the dispersion of swollen nanoplatelets or nanoclays into a solution of the monomer. Eventually, the monomers polymerize and penetrate the nanoplatelets or nanoclays, producing the PNC.

In the mechanical method, the polymer is dissolved using a co-solvent, and the nanofillers are dissolved using a different solvent to form a solution. Following a mechanical mixing of the two solutions, the solvent is removed, and nanoplatelets are intercalated by polymer chains. (Alexandre and Dubois 2000).

A subtype of intercalation method known as melt intercalation entails heating the polymer matrix at elevated temperatures in conjunction with the layered nanofillers. This process aims at achieving a uniform distribution of the nanofillers within the PNCs. This method is undeniably more ecologically friendly and superior to solution mixing because it eliminates the need for a solvent. It seamlessly integrates with modern industrial procedures such as extrusion and injection molding. A typical application is the production of regenerated PS/organoclay NCs through melt intercalation (Junior et al. 2014).

10.4.2 In Situ Polymerization Method

This is a highly effective method involving the swelling of the nanofillers in a solution of monomers. This allows the solution of the low-molecular-weight monomer solution to enter the nanofiller plates with ease, ensuring efficient and uniform polymerization through heat, radiation, or initiator. The PNCs synthesized by this method therefore have an intercalated or exfoliated microstructure.

In this method, the size of the clay or platelets is controlled using a polymer matrix as a template. The formation of clay layer in the presence of polymer chains is facilitated by *in situ* template synthesis. It is common practice to reflux the gel at an elevated temperature after dissolving the clay layers and polymer matrix in an aqueous solution, leading to the nucleation of clay layers interpenetrated by the polymer chains. However, it's important to consider that the heating process at elevated temperatures can cause the polymer to break. An example is the fabrication of exfoliated Nylon-6/clay nanocomposite in which the monomer solution of caprolactam containing uniformly distributed clay layers undergoes polymerization (Ray and Okamoto 2003).

10.4.3 Sol–Gel Method

The method comprises two related processes: sol and gel. The solution of monomers containing dispersed solid nanofillers is referred to as a "sol." The layers of the nanofiller are penetrated by the monomer solution, resulting in the formation of an interconnected, three-dimensional polymer network known as "gel." This process entails the polymer promoting the development of the nanofiller in a bottom-up approach (Tanahashi 2010). An example of this is the production of PS/SiO_2 NCs via the sol–gel process.

10.4.4 Direct Combination of the Polymer and Nanofillers

This innovative PNC fabrication method involves directly integrating the nanofillers with the polymer. Utilizing a top-down approach, this process effectively breaks down the agglomerated nanofillers throughout the mixing phase. The combination of polymers and nanofillers can be achieved through two strategies.

a. Melt Compounding: This method does not require any solvents. The agglomerated nanofillers uniformly disperse as individual NPs throughout the polymer matrix upon the introduction of the particles into the matrix above their glass transition temperature (T_g). This is accomplished by utilizing the shear stress generated by the viscous polymer melt.
b. The Solvent Method: In this method, a solvent is used for dissolving and breaking down the agglomerated nanofillers by sonication, while a co-solvent is used for dissolving the polymer. The solvent evaporates or coagulates, forming the PNC. Compared to melt compounding, the polymer matrix experiences reduced shear stresses in this method (Jouault et al. 2014).

10.5 ENVIRONMENTAL APPLICATIONS

Environmental pollution represents one of the most critical challenges of the current world that threatens ecosystems and wildlife. As industrial activities, urbanization, and unsustainable agricultural practices intensify, pollutants are continuously released into the environment, leading to widespread contamination, affecting climate patterns, biodiversity, and the overall quality of life. There are several pollutants, among which oil, heavy metals, particulate matter (PM), toxic gases, and organic dyes, which are particularly concerning due to their immense and persistent impacts. These pollutants, generated from various industrial, agricultural, and residential sources, contribute greatly to the deterioration of the quality of the air, water, and soil. PNCs are promising materials for degrading these pollutants to mitigate environmental pollution and help us keep a sustainable environment.

10.5.1 Monitoring Air and Water Quality

In a recent work, PNC-based sensor was designed to detect Hg(II) ions. Here, a synthetic polymer, poly(methyl methacrylate) (PMMA), and CuO NPs produced a potential PNC that combined with carbon nanofillers served as an effective sensor (Katowah et al. 2020). The sensor was used as an electrochemical sensor upon coating a glassy carbon electrode surface, and this exhibited good stability and fast response toward Hg(II) ions. A low-cost PNC was also designed by using polymelamine (PM) and NPs of graphitic carbon nitride (gC_3N_4) to detect cadmium and lead ions (Eswaran et al. 2021). The polymer and NPs were deposited electrochemically on the surface of the electrode to sense the metal ions from water samples. The developed PNC functioned as a ligand due to porous cavities provided by the NPs and the polymer provided amine functional groups that facilitated the formation of a complex with the metal ions. Excellent performance was exhibited in sensing lower concentration (0.1–1.0 μM) for the metal ions and even up to 0.02 μM for cadmium ions. Pesticides are another contaminant of water that poses serious health risks. Therefore, several attempts have been made to detect these contaminants to monitor water quality. In a recent work, a biosensor based on indigo carmine (IC) doped polypyrrole (PPy), dodecyl sulfate, and an acetylcholinesterase enzyme was developed to detect carbaryl pesticide (Loguercio et al. 2021). The enzyme facilitated high binding interaction of the pesticide with the sensor by adsorption mechanism. Furthermore, this biosensor exhibited good stability and good reproducibility.

H_2S is a toxic gas, often produced from sewage and industrial processes. In recent work, a new chemiresistive sensor using polyaniline (PAni)-based NC (PAni/WO_3/$CuCl_2$) was synthesized to detect this gas (Kumawat et al. 2022). The PAni/WO_3/$CuCl_2$ detected H_2S gas even at a low concentration (0.3 ppm). The incorporation of nano-scaled WO_3 favors effective detection by increasing the surface-to-volume ratio and p–n junction. A new PAni-based PNC exhibited excellent performance in NH_3 gas sensing applications with ultrafast response. In this work, a 3D framework based on nitrogen-based graphene was used as the supporting material for PAni/CuO NPs to produce a PNC that served as an effective sensor for NH_3 gas (Ahmadi et al. 2020). The interconnected graphene sheets increased the surface area and CuO NPs provided additional active sites. Also, the existing π-interactions in the PNC make it an effective adsorbent. This developed sensor has a fast response time of 30 seconds and can detect low concentrations between 50 ppb and 100 ppm.

10.5.2 Removal of Heavy Metals

Heavy metals, such as lead, mercury, and cadmium, tend to accumulate in living organisms. These metals disturb essential biochemical processes by binding to proteins and enzymes. They can cause severe damage to organs like the kidneys and liver, leading to chronic health conditions. Their persistence in the environment poses long-term health risks as they can enter into food and water supplies, leading to serious and irreversible damage. PNCs are advantageous for capturing heavy metals as they combine better adsorption properties, reusability, and often regeneration of polymers. In this context, a PNC based on a natural polymer, lignin has been synthesized that exhibited better performance in removing Pb(II) and Cu(II) from wastewater (Zhang et al. 2019). SiO_2-coated magnetic NPs combined with carboxymethylated lignin have been used as an adsorbent. This biodegradable adsorbent can adsorb Pb(II) and Cu(II) effectively within 30 seconds. Hydrogen bonding and ion exchange phenomena favor the PNC to function rapidly while adsorbing the heavy metals.

Graphene oxides (GOs) with high surface area often serve as excellent adsorbents due to the presence of several functional groups including carboxyl, epoxide, and hydroxyl groups. Their NCs also have numerous advantageous properties including tunable surface area and high mechanical strength. Furthermore, combining them with polymers enhances their versatility and adsorption capacities as the synergistic effect increases dispersion and interaction due to the additional functional groups provided by the polymers. Magnetic NPs (Fe_3O_4) are also a potential adsorbent as they serve many advantages, for example, ease of synthesis, rapid adsorption capacity, low toxicity, and affordability. Their great magnetic property favors reusability and ease of regeneration for PNCs. Therefore, polymer-assisted GO-based magnetic NCs have been used to remove Pb(II) and Cr(VI) from an aqueous solution (Manzoor et al. 2024a). Two PNCs, constituted of GO-based magnetic nanocomposite and chitosan (mGO/CS) as a natural polymer or polyaniline (mGO/PAni) as a synthetic polymer, were synthesized, which served as a potential adsorbent for Cr(VI) at pH 2 and Pb(II) at pH 5 from an aqueous solution. In this case, NH_2 and OH groups of the NC are protonated at a low pH of 2, creating a positively charged surface that draws the negatively charged hydrogen chromate ($HCrO_4^-$), which serves as a source of Cr(VI) ions (Ge and Ma 2015). Maximum adsorption of Pb(II) ion facilitates at pH 5, as the surface charge of the NCs allows the binding of Pb(II) ions with the NC over H^+ ions (Nadeem et al. 2016).

A PNC based on a synthetic polymer, PPy, was synthesized for the removal of Cr(VI) from wastewater (Sun et al. 2020). The PPy/Fe_3O_4/attapulgite composite can effectively adsorb Cr(VI) at a broad pH range from 2 to 6. Metal chelation, coulombic interactions, and reduction of the metal ion favor the adsorption phenomena. Here, the chemisorption exhibited the highest efficiency at a pH of 2. This PNC consists of Fe_3O_4, a magnetic NP that allows easy regeneration and reusability of the PNC. Magnetic NC based on PPy was also synthesized and investigated for the removal of Hg(II) from wastewater (Mashkoor et al. 2023). The magnetic NC is encapsulated by alginate, and together with PPy (Alg@Mag/PPy), these PNCs function as a potential adsorbent for Hg(II) ions. These Hg(II) ions are first adsorbed on the surface of PNC by the active groups and then penetrate the pores of the PNCs. Figure 10.3 represents the mechanism for the adsorption of Hg(II) ions by the PNCs. The binding affinity of the metal ion increases at elevated temperatures. The highest removal efficiency of the metal ion was obtained at 303 K, and the optimal pH was 7.

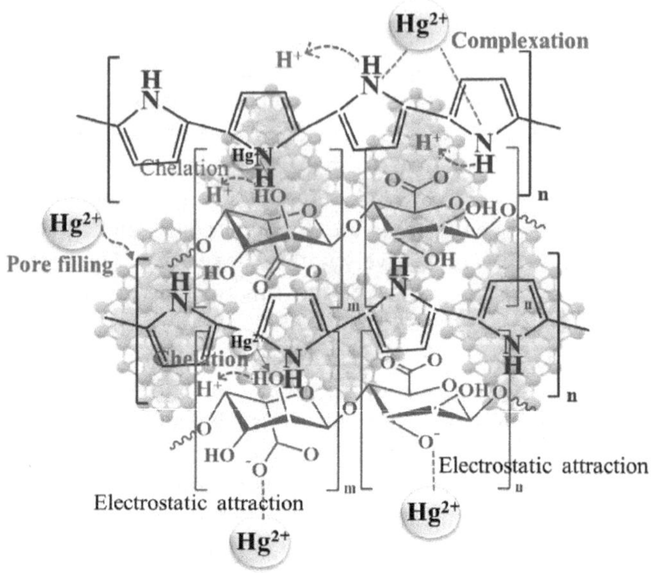

FIGURE 10.3 Mechanism for the removal of mercury(II) via Alg@Mag/PPy NCs (Mashkoor et al. 2023).

PAni-assisted magnetic core–shell NCs also exhibited excellent performance in removing Cr(VI) ions from wastewater (Lyu et al. 2019). Coral nanostructured PAni-coated Fe_3O_4 PNC (Fe_3O_4@CL-PAni) has proven to successfully adsorb harmful Cr(VI) ions. Here, the adsorption phenomenon that subsequently follows the reduction of Cr(VI) to Cr(III) is initiated by the electrostatic force of attraction between the protonated groups of the surface of PNCs and $HCrO_4^-$. Fe_3O_4 along with PAni provided many functional groups including imine and hydroxyl groups that favor faster reduction of Cr(VI). Also, less amount of Fe leach has been detected for the PNC, suggesting that PAni protected the Fe_3O_4 from corrosion.

10.5.3 Removal of Organic Dyes

Reactive red dye (RRD) is an organic anionic dye often found in textile wastewater. A biopolymer, sodium alginate (SA), and polyvinyl alcohol (PVA) based Cu-doped ZnO NC has proven to be an effective and selective adsorbent for RRD (Manzoor et al. 2024b). The PNC exhibited good porosity and adsorbed RRD by strong electrostatic force of attraction, as the polymers (PVA and SA) provide more functional groups with positive charge. The adsorption efficiency was maximum at pH 2, as the adsorbent surface contains more positive charges that serve as active sites for binding with the dye in the acidic condition. This adsorbent was regenerated and showed a high dye removal rate. Here, a simple mechanism was followed to remove the dye from the surface of the PNC by using 0.4 M NaOH. Also, PAni-based TiO_2 NC showed 98% removal efficiency for adsorbing methylene blue (Wang et al. 2019).

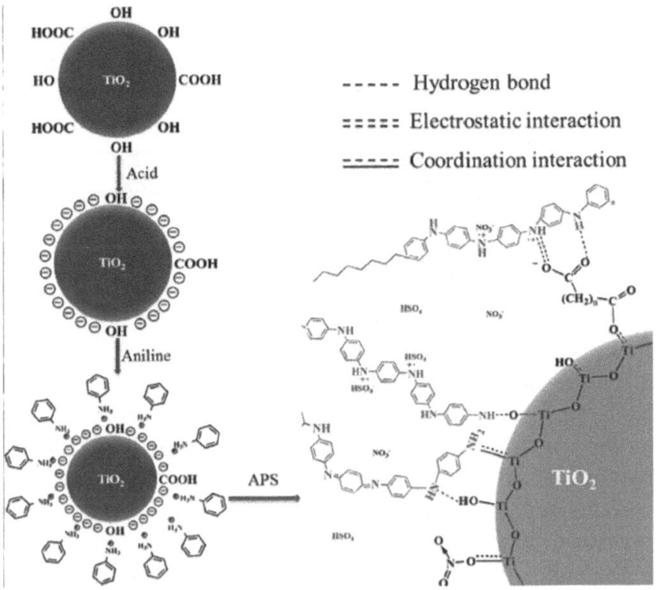

FIGURE 10.4 Schematic representation of the adsorption of methylene blue at PAni/TiO_2 NC surface (Wang et al. 2019).

Figure 10.4 presents the adsorption phenomena of PAni/TiO_2 NC. The primary mechanisms underlying the adsorption phenomena include chemisorption, hydrogen bonding, and intraparticle diffusions.

Chitosan is a notable biopolymer that is frequently used in wastewater treatment due to its capacity to remove contaminants by flocculation. Therefore, chitosan-based NCs have gained much attention as an excellent adsorbent. In this context, several attempts have been made to use this biodegradable PNC to remove organic dyes from wastewater. Chitosan-based magnetite (CS/Fe_3O_4) and graphene/chitosan/magnetite (Gr/CS/Fe_3O_4) NCs showed excellent photocatalytic degradation of several organic dyes. Gr/CS/Fe_3O_4 showed better performance and faster degradation compared to CS/Fe_3O_4 in degrading methyl orange, rhodamine B, methylene blue, and bromothymol blue (Maruthupandy et al. 2020). Gr/CS/Fe_3O_4 exhibited 100% degradation of rhodamine B within 40 minutes while CS/Fe_3O_4 took up to 90 minutes. Gr/CS/Fe_3O_4, with CS/Fe_3O_4 NCs placed on each side of the graphene sheets, have increased surface area. Furthermore, Fe(II) ion of the NCs captured dye species that increased the rate of degradation. The conductive bond between graphene sheets and CS/Fe_3O_4 NCs delayed the recombination of the electron–hole pairs, as chitosan and graphene functioned as electron reservoirs to capture electrons from magnetite NPs under irradiation of visible light. PNC based on modified gelatin and montmorillonite clay NC was also synthesized to remove malachite green from wastewater (Abu et al. 2024). This PNC showed effective adsorption of the dye followed by electrostatic force of attraction and hydrogen bonding. The adsorbent possesses good recyclability and represents it as one of the cost-effective water-purifying adsorbents.

10.5.4 Removal of Toxic Gases

A porous membrane based on a PNC has been developed to capture CO_2 from the air. A synthetic polymer, polyvinylidene fluoride (PVDF) with hexafluoropropylene (HFP), and amino-silica NPs produced PVDF-HFP/SiO_2 that exhibited high mechanical properties and high surface area for capturing CO_2 (Khdary and Abdelsalam 2020). This is a low-cost PNC with high stability that functions as an efficient membrane. A highly selective membrane was also designed based on a PNC to capture CO_2 gas molecules. The PNC is based on a commercial polymer, polyetherimide (PEI), and functionalized GO nanosheets (Koolivand et al. 2014). The GO was functionalized by ethylenediamine and polyethylene glycol (PEG) which facilitated high selectivity for CO_2 by the membrane. The development of free volumes surrounding the layered structure of functionalized GO nanosheets was made possible by the strong interfacial interaction between the PEI matrix and the functionalized GO. Consequently, the free volume distribution leads to effective CO_2 separation. The PNCs have also been used to capture H_2S gas, which is also toxic and often contributes to acid rain. Polystyrene-based iron oxide NC was synthesized for capturing H_2S gas (Blatt et al. 2014). The PNC served as a mesoporous system that favors the capturing of H_2S gas molecules. This exhibited significantly higher adsorption capacity than other adsorbents and the adsorption phenomena can occur at room temperature.

10.6 BIOMEDICAL APPLICATIONS

PNCs have also emerged as promising materials in biomedical applications. Their exceptional and tunable properties with superior biocompatibility make them a potential alternative to traditional polymeric materials to address complex medical needs. Modern healthcare systems still face the major challenge of replacing damaged organs with prostheses or mechanical devices. Since they can't restore normal bioactivities and respond accordingly. Also, their long-term implantation can affect the normal functions of the surrounding cells in the body. PNCs can be advantageous, as they have superior biocompatibility, tunable properties, and higher efficiency. Tissue engineering supports the key strategy that involves developing scaffold biomaterials that can mimic native cell and tissue structures to promote cell growth. Several natural polymers and NPs promote bio-functionality, proliferation of cells, and biocompatibility. Figure 10.5 represents the versatile properties of Au NPs that enhance cell adhesion, promote the proliferation of stem cells, and promote maturation in tissue engineering.

In the PNC system, the synergistic effect produced by combining polymers with NPs enables them to create innovative and more efficient scaffolds for some regenerating tissues such as skin, nerves, and bones. In addition, their capacity to regulate the release of therapeutic agents makes them indispensable for targeted drug delivery systems. The porous and functionalized surfaces enhanced the efficiency of the treatment with reduced side effects. Moreover, the PNC-based materials offer superior physicochemical and magnetic properties that are advantageous for developing advanced medical devices. The most prevalent applications of PNCs in wound healing, bone and neural regeneration, and cancer diagnosis and treatment are discussed below.

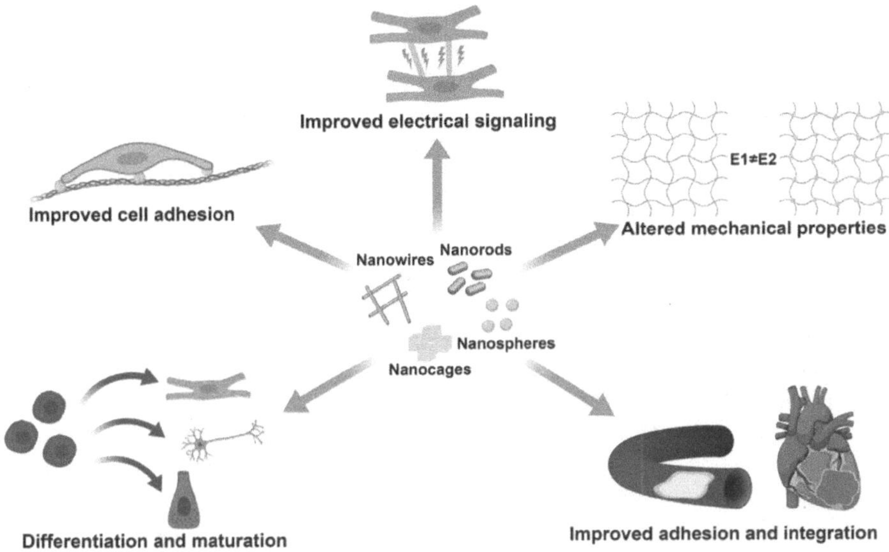

FIGURE 10.5 Schematic representation of functions of Au NPs in tissue engineering. (Yadid et al. 2019).

10.6.1 Bone and Neural Regeneration

PNCs are useful in applications involving bone and neural regeneration since they can mimic the structure and characteristics of the extracellular matrix found in local nerve cells and bones. They can also promote cell development by reproducing biochemical properties that can encourage the formation of new tissues. Conductive polymers are promising materials to generate NCs for the regeneration of bone and tissue. They have good electrical conductivity and electroactive properties that stimulate bone growth and repair. Also, the conductive polymers interact with the cell that facilitates bone regeneration. They can produce electrical signals to promote and maintain cell activities. PNCs can be tailored to match the requirements of the patient and several attempts have been made to prepare PNC-based materials to design effective implants. For instance, a PNC based on simvastatin and NPs of TiO_2/PAni contained in polycaprolactone (PCL) was synthesized (Rezk et al. 2020). The prepared PNC facilitated the release of simvastatin in a controlled way that further stimulated osteogenesis. The composite system also proved to be biocompatible with favorable morphology that matches the requirements for bone applications.

In the neural regeneration application, several PNCs exhibited better performance by imitating native cells. For example, TiO_2 nanotubes have antimicrobial properties and coating them with a highly conducting polymer (PAni) produces an effective NC system (Shrestha et al. 2019). The addition of the polymer was attributed to decreased resistance for charge transfer with enhanced antibacterial properties. The prepared system successfully stimulated nerve cells and showed notable achievement in performing cellular functions. Modified PPy NPs fabricated onto poly-L-lactic acid (PLLA) resulted in high stability and conductivity of the prepared material (Shang et al. 2019).

PPy modified with GO and sodium dodecyl benzenesulfonate showed enhanced cytocompatibility and high stability. In addition, in vivo experiments revealed that the prepared material can be used for nerve repair. An efficient conductive scaffold has been developed by doping hyaluronic acid with poly(3,4-ethylenedioxythiophene) NPs and incorporating it into the polymeric matrix of chitosan/gelatin (Wang et al. 2017). The prepared scaffold served as a substrate for supporting synapse growth and cell proliferation facilitating its application as nerve tissue regeneration.

10.6.2 Wound Healing

Several NPs have antimicrobial properties and incorporating them into the polymer matrix increases their acceptability in wound-healing applications. Moreover, conductive PNCs hold immense potential to support wound healing by controlling bleeding, stimulating cell regeneration through bioactive delivery, responsive healing, and preventing infection (Khan et al. 2021). PCL combined with chitosan-graft-PAni produced a new nanofibrous membrane of great mechanical properties and multifunctionality (He et al. 2020). The prepared membranes were used as a wound dressing that showed antibacterial properties, electroactivity, scavenging capacity, and cell proliferation. In addition, the membrane showed collagen deposition and higher angiogenesis properties that supported skin repair. Researchers have designed a new potential wound-healing tool for diabetic patients by incorporating PLX/ZnO NCs with the extract of *Centella asiatica* (Wang et al. 2024). Poloxamer (PLX) is a block copolymer that exhibits self-assembly, biocompatibility, and versatility, which facilitate the multifunctionality of the prepared material. Here, ZnO and extract of *Centella asiatica* were encapsulated in the polymer matrix. The PNC showed great bactericidal inhibition, fast self-healing, increased proliferation of cells, and biocompatibility that support wound healing of a diabetic patient to avoid foot ulcers. ZnO and chitosan produced a PNC to use as an effective tool for supporting the healing mechanism of an infected wound (Shu et al. 2023). The prepared material exhibited antibacterial and antioxidant properties. In addition, the material showed biocompatibility without any cytotoxic effect, reduced inflammation, inhibited bacterial growth, and enhanced collagen deposition.

10.6.3 Diagnosis and Treatment of Cancer

Since biopolymer-based NCs have improved biocompatibility, they have drawn a lot of interest in the treatment of cancer. In this context, a pH-sensitive PNC was prepared by combining a biopolymer, chitosan with magnetite and hydroxyapatite NPs (Bakhshi et al. 2024). An anticancer agent, doxorubicin was loaded in the prepared NC that exhibited a higher release rate (around 80%) at pH 5 while it was very low at pH 7.4. Moreover, the prepared NCs can travel toward a targeted cell as a response to the external magnetic field. The material also showed effective toxicity to be used as a tool for destroying uterine cancer.

Laser irradiation often causes healthy tissue damage while detecting cancer cells. To reduce the damage, several photothermal therapies of low power can be used as an alternative. Therefore, polydopamine (PDA), CDs, and hollow structured mesoporous carbon produced a new drug delivery system with multifunctionality. The prepared NC (HMC-SS-PDA@CDs) acted as a photothermal responsive system (Lu et al. 2021). CDs and PDA are pH-sensitive and NIR-adsorbing agents. The mesoporous carbon structure is connected with PDA and CDs through disulfide bonds. These combined properties prove the multifunctionality of the system as it is sensitive and creates responses with the release of pH, NIR, and glutathione (GSH). Porous carbon can act as a fluorescence quencher, but the interaction between PDA and the carbon structure leads to the red-shift in the fluorescence emission wavelength by the fluorescence resonance energy transfer effect. Consequently, this phenomenon resulted in the photothermal response created by the system. Additionally, the porous carbon structure provided a hollow frame that facilitated a high loading capacity for doxorubicin hydrochloride (DOX) that was incorporated into the NC, which favored enhanced drug release and great inhibition rate for tumor cells. Furthermore, it performed effective thermo-chemotherapy even at low-power irradiation in the antitumor experiments.

CD is another promising material to use as a nanofiber to prepare PNCs for the treatment of cancer cells. A stimuli-responsive biosensor was developed by combining hydrogels of ureidopyriminone-conjugated gelatin and CDs containing diselenide groups (Won et al. 2020). The sensor can detect cancer cells under the irradiation of some reactive species of oxygen or GSH, as they break down the bonds of the diselenide groups. This results in the degradation of the CDs, and consequently, it affects the electrochemical signals that facilitate the successful detection of cancer cells. The irradiation of the reactive species and GSH increases conductivity, adhesiveness, and self-healing properties and also favors the formation of intermolecular interactions via hydrogen bonding. The material was biocompatible with no inflammation and showed fast self-healing properties in the presence of MDA-MB-231 cells.

10.7 ENERGY STORAGE AND APPLICATIONS

PNCs have been gaining attention in energy-related applications for the past few decades. Integration of CNTs or graphene into the polymer matrices significantly boosts electrical conductivity. In energy storage applications, these materials are often used as advanced electrode materials offering higher charge storage capacities and faster charging–discharging rates. They can contribute to sustainability by improving the efficiency and lifespan of batteries, supercapacitors, fuel cells, etc. Moreover, their unique properties facilitate their use in smart energy technologies as well as meeting the continuously increasing demands of the energy sector to meet the global energy crisis.

10.7.1 SUPERCAPACITOR

Supercapacitor is a high-capacity energy storage device, and introducing PNCs in a supercapacitor can significantly enhance mechanical properties, electrical conductivity, and flexibility. Incorporation of NPs can significantly minimize dielectric loss and facilitate their use in the supercapacitor. For instance, PAni-based NiO NC was prepared by the oxidative polymerization method. The PNC showed low dielectric loss and proved its potential for use as an effective dielectric material (Roy et al. 2018a).

Several metal oxide-based PNCs were developed to introduce them into energy storage devices. For example, V_2O_5-incorporated PAni NC was prepared for fabricating electrode materials. The metal oxide provided the higher oxidation state of the metal ion that facilitated enhanced electron storage for the fabricated electrode. The uniform dispersion of the metal oxide resulted in a porous structured PNC that enhanced the electrochemical behaviors of the electrode. Since there was less resistance to charge transfer, it also demonstrated a higher capacitive retention rate with high power and energy density (Thakur et al. 2024). A mixed metal oxide was also used to prepare PVP- and PVA-based NC to fabricate the electrode of a supercapacitor device (Murugan et al. 2024). PVP/PVA/$CoMoO_4$ was developed to fabricate the cathode, and activated carbon was used as the anode. $CoMoO_4$ combined with the polymers produced a synergistic effect that facilitated increased power and energy density properties of the device. The device exhibited high specific capacitances with reduced internal resistance and high conductivity. This solid-state supercapacitor provided a better charging–discharging capacity with higher stability than other conventional supercapacitors.

10.7.2 Fuel Cells

Fuel cells are widely recognized for their high efficiency, fuel flexibility, and versatility. Nafion® membranes are the most used material in fuel cells though it is very expensive. There has been a surge of interest to develop low-cost membranes as a replacement for this expensive membrane. For instance, metal oxide-doped PNC was used to develop a membrane to replace the Nafion® membrane. Here, a polymer blend of PVA, PVP, and poly (ethylene oxide) (PEO) was doped with (SO_4–TiO_2) and (PO_4–TiO_2) nanotubes to prepare an ion exchange membrane for direct borohydride fuel cell. Because of the TiO_2-based nanotubes, the membrane had improved oxidative and thermal stability, reduced degree of swelling, and a greater capacity for ion exchange. It also exhibited great power density similar to the Nafion®117 membrane. Hence, the prepared PNC proved its potential as a replacement for the expensive fuel cells (Gouda et al. 2020). A quaternized nanocellulose was also incorporated into the quaternized PVA matrix to develop an anion exchange membrane. The nanocellulose was extracted from the fruit bunch of oil palm trees. The membrane showed higher ion conductivity since the addition of the nanocellulose reduced the loss of ion conductivity with alkaline stability of the prepared membrane. The content of ammonium group and the dimensional stability of the membrane both were greatly improved with the addition of larger nanocellulose content. It also showed excellent selectivity for anions and proved to be a potential membrane for use in fuel cell applications (Yunphuttha et al. 2024).

10.7.3 Solar Cells

Solar cells are the most popular energy-producing devices. Numerous efforts have been made to use PNCs in solar cells to improve their efficiency and versatility. For instance, a PVDF-based film was developed by incorporating a mixed metal oxide/CNT NC in the polymer matrix. Nanopowder of $SrTiO_3$ and CNTs produced NC

($SrTiO_3$/CNT), which was further incorporated into the polymeric matrix of PVDF. The film showed enhanced optical properties, spherulite morphology, and modified electronic properties. Moreover, 0.1 wt% addition of the NC facilitated increased refractive index, polarizability, and higher bandgap of the prepared film. The prepared film holds immense potential for use in solar cells (Taha et al. 2024).

SiO_2 NPs have tunable structures, characteristic bandgaps, and enhanced dielectric properties. Therefore, they are often used to fabricate solar cells for better performance. For example, a ternary PNC was designed by incorporating SiO_2 NPs. The system consists of three polymers: poly(acrylamide) (PAAm), PVA, and poly(vinyl pyrrolidone) (PVP). The material showed increased electrical conductivity, improved dielectric constant, and energy gap. The addition of SiO_2 NPs significantly improved energy gap from 4.8 to 3.4 eV and enhanced antibacterial activity toward *E. coli* and *S. aureus* (Alawi and Al-Bermany 2023).

10.8 INDUSTRIAL APPLICATIONS

The advancement of PNCs has revolutionized numerous fields, including industrial applications. Industries are continuously searching for innovative solutions to meet their demands such as excellent performance, sustainability, durability, and cost efficiency. As the PNCs exhibit enhanced mechanical properties, improved thermal and electrical conductivity, higher stability, lightweight, and versatility, they are considered promising materials for achieving these goals. PNCs have reached almost every industrial sector from cosmetics, food packaging, and automotive industries to aerospace applications. But for industrial applications requiring large-scale production, scientists are still attempting to improve the properties of NCs.

In the automotive industry, the most challenging task is to reduce weight and decrease the energy consumption of the vehicles. Introducing PNCs would provide a solution for the problem, and materials of this kind can significantly reduce the weight of the vehicles and enhance their properties for better functionality (Harun-Ur-Rashid et al. 2023). In the packaging industries, PNCs are continuously gaining attention due to their antimicrobial property and other advantages. For instance, Ag NPs were incorporated into the matrix of plasma-polarized hexamethyldisiloxane monomer to prepare a PNC that was used for coating the cotton fabric to develop hydrophobic medical wear. The transparency of the coated fiber can be modified by the deposition of the Ag NPs, as reduced deposition and lower concentrations of the NPs facilitated the formation of smaller-sized particles that resulted in better transparency of the coated fabrics. The NC thin film exhibited great antibacterial activity against a Gram-positive bacterium (*Staphylococcus epidermidis*) even after ten washing cycles. In addition, the fabric showed good optical properties and high hydrophobicity (Irfan et al. 2019). Water-repellent fabric was developed by coating a new PNC on the cotton fabrics. The PNC was prepared by silica NPs with two nontoxic polymers, the acrylic copolymer of C-6 perfluorinated, and acetoxy-cure silicone. The treated cotton fabrics showed high hydrophobicity with low water penetration ability. It also showed excellent resistance to wear abrasion up to 30 cycles of the abrasion test and offered a porous structure to facilitate the breathability of the fabrics (Zahid et al. 2017).

In the packaging industries, PNCs are continuously gaining attention due to their antimicrobial property and other advantages. Ag NPs with antimicrobial properties can significantly increase the antimicrobial activities of PNC-based materials. Ag/PAni NCs, prepared using w/o microemulsion, showed higher thermal stability compared to the constituent polymer (Fatema et al. 2018). It also showed excellent antimicrobial properties against *E. coli* and *S. aureus*.

Cotton fabrics were treated with silver NPs, chitosan, and an extract of date seed to improve their functional properties (Ramadan et al. 2020). The extract contains antioxidants, flavonoids, and alkaloids that possess inhibitory properties against microbes. Chitosan has numerous charged groups in its backbone, which can interact with the cell wall of bacteria and inhibit bacterial growth. Silver NPs also possess high antimicrobial properties, as they contain high surface area that allows higher interaction with the microorganisms. They also showed significant inhibition of the growth of bacteria by producing oxidative stress and damaging their intracellular biomolecules. Therefore, this fabric showed excellent antimicrobial activity against some microorganisms due to the combined antimicrobial properties of date seed extract and silver NPs. The treated fabric also showed increased water absorbance and air permeability. Therefore, the treated fabric can be used in food packaging, for example, seeds or other powder materials.

The use of biopolymer-based NCs as cosmetic carriers attracted much attention among researchers. An NC of layer-structured double hydroxide and chitosan was prepared for the selective release of a cosmetic vesicle, liposome capsules (Kim et al. 2021). The prepared PNC gained a porous structure and anion selectivity. The porous structure provided long-term storage and improved facile uptake. Adjusting pH and anions in the solution governs the selective release of retinol-containing liposomes. The ionic interaction between the nanosheets of double hydroxide and the polymer network of chitosan facilitated the effective blocking and controlled release of the liposome capsules. The PNCs showed good reusability and notable biocompatibility, also their excellent properties allow slow release and fast uptake of the cosmetic vesicles proving it a suitable carrier for cosmetics.

PNCs exhibited excellent performance in producing ultraviolet (UV) shielding materials. MnO_2 NPs were incorporated into the PVA matrix to produce PVA/MnO_2 NC film by solution casting techniques (Roy et al. 2018b). The carbonyl and hydroxyl groups in the backbone of PVA formed a complex with the MnO_2 NPs. This facilitated the excellent absorption of UV radiation as the vibration of the two functional groups shifted and exhibited a less intense band. The prepared PNC film entirely shielded UV radiation and proved itself a potential material for use in healthcare, automotive, packaging, cosmetics, textiles, electronics, and aerospace engineering.

10.9 POLYMER NANOCOMPOSITES AS SMART MATERIALS

"Smart materials" can constantly detect and respond to different forms of physical or chemical stimuli and then adjust their properties accordingly (Hsu et al. 2011). Physical stimuli encompass magnetic fields, light, electric fields, and temperature, while chemical stimuli include chemical recognition, enzymatic recognition, pH, and solvent (Hsu et al. 2011).

Stimuli-responsive PNCs are a cutting-edge form of smart materials, uniquely tailored to exhibit specific responses to stimuli based on the interplay between the polymer and the type of nanofiller embedded within. The intensity of stimuli can effectively trigger and control the response rate of the smart PNCs. Smart polymeric materials and PNCs are utilized in catalysis, drug delivery, actuators, sensors, prosthetics, decontamination, automotive, biomimetic developments, and aerospace (Cardoso et al. 2018).

10.9.1 Thermoresponsive

Thermoresponsive polymer matrices are smart materials and can modify the structure related to hydrogen bonding due to temperature changes. One excellent example of a thermoresponsive polymer is the hydrogel, which undergoes a significant change in volume as the temperature varies. This change occurs through a reversible transition of the polymer chain from a hydrolyzed condition to a precipitated condition (Vera et al. 2020). Nanofillers can enhance the capabilities of drug release, mechanical and other properties, broadening the potential of PNCs as replacements for biological substrates, living organs or tissues, and mechanical actuators (Yue et al. 2019). Examples of thermoresponsive PNCs include PE/CNT, PP/CNT, polyimide/CNT, and polyoxymethylene/metal oxide, etc. (Santos et al. 2016).

10.9.2 Magnetoresponsive

By incorporating magnetic NPs, magnetoresponsive PNCs can respond to an applied magnetic field by changing their shape or movement. This increases the sensitivity of the PNCs to magnetic fields. The combination of magnetic nanofillers with thermoresponsive polymers allows for structural changes via localized heating from an alternating magnetic field (AMF). The Neel relaxation process and Brownian motion drive these changes as the AMF prompts the nanofillers to release heat through vibration (Van Berkum et al. 2013). PU/Fe_2O_3 and PLLA/Fe_2O_3 NCs are examples of magnetoresponsive PNCs. They can be applied as shape memory composites (SMC), shape memory polymers (SMP), and artificial muscles (Soto et al. 2018, Zhao et al. 2019).

10.9.3 Electroactive

Electric currents can change the shape or size of SMP or SMC materials by generating localized heat through the Joule heating effect, enabling the generation of composites with conducting properties. Inactive nanofillers combined with an intrinsically conducting polymer matrix can produce an electroactive PNC with improved mechanical properties. However, using active nanofillers in a nonconducting polymer matrix may be a better option. Nanofillers such as graphene, GO, and CNTs are good choices as active ingredients; however, their nonhomogeneous dispersion in polymer matrix restricts their application (Lu et al. 2019). The properties of conducting nanofiller and conducting polymer can also be combined to achieve superior electroactive PNCs. An example of an electroactive PNC is the reduced graphene oxide (rGO) nanofiller dispersed in acrylamide (AAm) and 2-acrylamide-2-methyl propane sulfonic acid (AMPS) through *in situ* polymerization to produce an rGO/poly(AMPS-co-AAm) nanocomposite, which can swell or dwell with the application of an electric field (Yang et al. 2017).

10.9.4 Photosensitive

Photosensitive PNCs can be synthesized by employing photosensitive nanofillers or photosensitive polymer matrices. Photosensitive nanofillers, when combined with polymers that are not sensitive to light, can form photosensitive PNC due to localized surface plasmon resonance (LSPR) of NPs (Alebrahim et al. 2024). The LSPR effect stimulates NPs to interact with photons, emitting potent electromagnetic radiation and facilitating rapid heat transfer from the NPs to the surrounding polymer matrix causing them to respond. Photosensitive PNCs show significant promise for various applications in the field of drug delivery and cancer therapy using nanocarriers.

Photosensitive polymer matrices, which contain chemical chromophore groups such as diaryl ethane, coumarin, or azobenzene (Gafiullina et al. 2022), can also be used to synthesize photosensitive PNCs. These can change polarity, chirality, conjugation, charge, or amphiphilicity due to light sensitivity.

10.9.5 pH-Sensitive

pH-sensitive polymers or PNCs contain tiny moieties capable of accepting or donating protons due to change in pH. Liu et al. have successfully produced pH-sensitive graphene-based PNCs using pyrene-terminated poly(2-*N*-(dimethyl amino ethyl acrylate) (PDMAEA) and poly(acrylic acid) (PAA) as the matrix (Liu et al. 2010). These NCs display changes in solubility in both aqueous and organic systems in response to pH changes, demonstrating a high degree of responsiveness. Polymeric nanocarriers that are responsive to changes in pH and redox potential have significant prospects as advanced biomaterials for selective drug delivery during effective tumor treatment (Gaddimath et al. 2024).

10.9.6 Biointeractive

Biointeractive PNCs are innovative smart materials that result from combining polymers with nanofillers to prepare composites that interact with biological systems. These advanced materials are engineered to possess unique properties and functionalities that are highly beneficial in biomedical and environmental applications. They have diverse applications in drug delivery systems, medical implants, and tissue engineering. Notably, some biointeractive PNCs are designed to be biodegradable, making them environmentally friendly. Another significant development is the enzyme-responsive smart PNCs, which have the potential to bring about revolutionary advancements in biomaterials for cell supports, injectable scaffolds, and drug delivery (Lavrador et al. 2021).

10.9.7 Self-Healing

They are typically composed of specifically engineered polymeric substances with self-repairing capabilities in response to various stimuli. Functional groups onto the framework of the polymer allow the polymeric compounds to undergo self-healing. These functionalized polymers play a crucial role as matrices in PNCs, imparting remarkable self-healing properties. The self-healing process within the polymer can be intrinsic, where the polymer inherently possesses healing properties, or extrinsic, where external healing agents are incorporated. Self-healing PNCs involve processes such as shape recovery, crack healing, and dimensional stability, which are beneficial for extending the lifespan of objects and preventing premature disposal (Orellana et al. 2021).

10.10 POLYMER NANOCOMPOSITES FOR SUSTAINABLE FUTURE

In an era marked by escalating environmental concerns and the urgent quest for sustainability, innovative solutions are leading the way in the field of material science. Among the most promising advancements is the development of PNCs, an interdisciplinary innovation merging polymer with nanoscale materials to produce composites with extraordinary properties. This technological leap offers a transformative potential for various industries, from construction to electronics, aligning closely with global objectives for sustainable development. Recent technological advancements have significantly reduced the size of electronic devices, making them lighter, slimmer, and more portable. However, as production ramps up, so does electronic waste. In response, the scientific community is intensifying research to develop a new class of materials designed for biodegradation after use. These innovative materials are not only sustainable but also exhibit biocompatibility during their decomposition in the natural environment (Kim et al. 2019). This research is pushing the boundaries of material science and chemistry. Research on polymer-graphene NCs has concentrated on their potential in various electronic applications, including sensors, printable conductors, supercapacitors, electroluminescent devices, actuators, batteries, light-emitting diodes (LEDs), memory devices, and transistors (Kausar 2016, Sandler et al. 2003).

The physical, chemical, and mechanical properties of conventional polymers are enhanced by integrating NPs into polymer matrices. These enhancements often translate into more efficient use of materials, improved durability, and reduced environmental impact. In the context of energy, these composites are particularly valuable. Their unique properties can lead to breakthroughs in energy storage, such as advanced batteries and supercapacitors, as well as more efficient energy conversion systems, including photovoltaic cells. Hosseini et al. developed CS/GM/Fe_3O_4/PAni NCs for supercapacitor applications by integrating Fe_3O_4 NPs into chitosan/graphene oxide-multiwall CNTs (CS/GO MWCNTs) and subsequently grafting them with PAni through *in situ* polymerization. The resulting NCs exhibited an impressive specific capacitance of 1513.4 F/g at a current density of 4 A/g, which is 1.9 times greater than that of the control electrode CS/GO MWCNTs/Fe_3O_4 without PAni (Hosseini et al. 2021).

Renewable energy technologies benefit significantly from the application of PNCs. For instance, the integration of nanomaterials into photovoltaic cells can improve their efficiency, making solar energy more viable as a mainstream energy source. Graphene/polymer NCs are used as transparent conductive electrodes, active layers, and interfacial layers in solar cells (Díez-Pascual et al. 2018). Similarly, in the realm of energy storage, these composites contribute to the development of lighter, more efficient batteries that support the transition from fossil fuels to renewable energy sources. PNCs play a crucial role in lithium-ion batteries (LIBs). They have been the most important and widely utilized rechargeable batteries because of their high operating voltage, low toxicity, substantial capacity, and extended cycle life (Ding et al. 2014). PNCs are essential for reducing dependency on nonrenewable resources and mitigating the effects of climate change since they prolong the lifespan and performance of these technologies.

Moreover, PNCs offer exciting possibilities for green manufacturing processes. Their ability to be tailored for specific applications means that less material can be used to achieve the same or even superior performance compared to traditional materials. This reduction in material usage, coupled with the potential for recycling and reusability, positions PNCs as a cornerstone in the development of sustainable materials.

As the world moves toward a greener future, the continued research and application of PNCs will be pivotal. Their potential to revolutionize various sectors while contributing to energy efficiency and renewable energy solutions highlights their importance in creating a sustainable future. This exploration of PNCs not only addresses the immediate needs of contemporary technology but also aligns with the long-term vision of environmental stewardship and conservation of resources.

10.11 CONCLUSIONS

The advancements in polymer-based nanotechnology unlocked new possibilities in numerous fields, including healthcare, agriculture, energy, and environmental conservation, by exploring the excellent properties of materials at the nanoscale. PNC-based materials offer eco-friendliness, cost-effectiveness, and multifunctionality to meet the demands of modern civilization. Polymers with several advantageous NPs produce a synergistic effect that makes the material appealing to researchers and industries. Their unique physicochemical properties offer a promising solution to mitigate environmental pollution for keeping a sustainable future for the mankind. In this chapter, we have discussed how PNCs are driving progress in different fields and affect our everyday lives. In recent years, PNC materials entered into every sector of our lives. In addition, smart polymer-based materials enhance the functionality and performance of everyday products, contributing to a more responsive and adaptive lifestyle. Furthermore, their attractive features make them indispensable for shaping a more intelligent and responsive future.

Although NCs have immense potential, there are still some challenges that must be overcome before enhancing their application in large-scale production. Further research strategies are essential to optimize the techniques for the synthesis that will increase their biocompatibility, reusability, and efficiency. The continuous effort can contribute to produce an effective solution for mitigating environmental pollution while meeting the demands of the growing civilization. Also, the continuous exploration of nanotechnology can ensure the decisive role played by PNCs in shaping the materials of tomorrow. In summary, PNCs lead to the nano revolution by merging nanotechnology with polymer science to create versatile materials that can encompass our daily lives and shape future technological landscapes.

ACKNOWLEDGMENTS

The authors acknowledge the support from the Student Promotion and Support Unit of the University of Dhaka for DUNC. MABHS also acknowledges the Ministry of Science and Technology, Government of Bangladesh for research grant under Special Allocation for Science and Technology.

REFERENCES

Abu Elella, M. H., Aamer, N., Abdallah, H. M., López-Maldonado, E. A., Mohamed, Y. M. A., El Nazer, H. A., Mohamed, R. R. 2024. Novel high-efficient adsorbent based on modified gelatin/montmorillonite nanocomposite for removal of malachite green dye. *Sci. Rep.*, 14(1), 1228.

Ahmadi Tabar, F., Nikfarjam, A., Tavakoli, N., Nasrollah Gavgani, J., Mahyari, M., Hosseini, S. G. 2020. Chemical-resistant ammonia sensor based on polyaniline/CuO nanoparticles supported on three-dimensional nitrogen-doped graphene-based framework nanocomposites. *Microchim. Acta*, 187, 1–13.

Alawi, A. I., Al-Bermany, E. 2023. Newly fabricated ternary PAAm-PVA-PVP blend polymer Doped by SiO_2: Absorption and dielectric characteristics for solar cell applications and antibacterial activity. *Silicon*, 15(13), 5773–5789.

Alebrahim, M. A., Ahmad, A. A., Migdadi, A. B., Al-Bataineh, Q. M. 2024. Localize surface plasmon resonance of gold nanoparticles and their effect on the polyethylene oxide nanocomposite films. *Phys. Rev. B Condens. Matter*, 679, 415805.

Alexandre, M., Dubois, P. 2000. Polymer-layered silicate nanocomposites: Preparation, properties and uses of a new class of materials. *Mater. Sci. Eng. R Rep.*, 28(1–2), 1–63.

Armentano, I., Dottori, M., Fortunati, E., Mattioli, S., Kenny, J. M. 2010. Biodegradable polymer matrix nanocomposites for tissue engineering: A review. *Polym. Degrad. Stab.*, 95(11), 2126–2146.

Bakhshi, V., Poursadegh, H., Amini-Fazl, M. S., Salari, D., Javanbakht, S. 2024. Synthesis and characterization of bio-nanocomposite hydrogel beads based on magnetic hydroxyapatite and chitosan: A pH-sensitive drug delivery system for potential implantable anticancer platform. *Polym. Bull.*, 81(8), 7499–7518.

Blatt, O., Helmich, M., Steuten, B., Hardt, S., Bathen, D., Wiggers, H. 2014. Iron oxide/polymer-based nanocomposite material for hydrogen sulfide adsorption applications. *Chem. Eng. Technol.*, 37(11), 1938–1944.

Cardoso, V. F., Correia, D. M., Ribeiro, C., Fernandes, M. M., Lanceros-Méndez, S. 2018. Fluorinated polymers as smart materials for advanced biomedical applications. *Polymers*, 10(2), 161.

Coleman, J. N., Khan, U., Blau, W. J., Gun'ko, Y. K. 2006. Small but strong: A review of the mechanical properties of carbon nanotube-polymer composites. *Carbon*, 44(9), 1624–1652.

Cruz-Chú, E. R., Villegas-Rodríguez, G. J., Jäger, T., Valentini, L., Pugno, N. M., Gkagkas, K., Gräter, F. 2020. Mechanical characterization and induced crystallization in nanocomposites of thermoplastics and carbon nanotubes. *Npj Comput. Mater.*, 6(1), 151.

Díez-Pascual, A. M., Luceño Sánchez, J. A., Peña Capilla, R., García Díaz, P. 2018. Recent developments in graphene/polymer nanocomposites for application in polymer solar cells. *Polymers*, 10(2), 217.

Ding, K., Gu, H., Zheng, C., Liu, L., Liu, L., Yan, X., Guo, Z. 2014. Octagonal prism shaped lithium iron phosphate composite particles as positive electrode materials for rechargeable lithium-ion battery. *Electrochim. Acta.*, 146, 585–590.

Doagou-Rad, S., Islam, A., Merca, T. D. 2020. An application-oriented roadmap to select polymeric nanocomposites for advanced applications: A review. *Polym. Compos.*, 41(4), 1153–1189.

Eswaran, M., Tsai, P. C., Wu, M. T., Ponnusamy, V. K. 2021. Novel nano-engineered environmental sensor based on polymelamine/graphitic-carbon nitride nanohybrid material for sensitive and simultaneous monitoring of toxic heavy metals. *J. Hazard. Mater.*, 418, 126267.

Fatema, U. K., Rahman, M. M., Islam, M. R., Mollah, M. Y. A., Susan, M. A. B. H. 2018. Nanocomposites of polyaniline with silver nanoparticles prepared using water in oil microemulsions as antibacterial agents. *Macromol. Symp.*, 379(1), 1700031.

Fu, S., Sun, Z., Huang, P., Li, Y., Hu, N. 2019. Some basic aspects of polymer nanocomposites: A critical review. *Nano Mater. Sci.*, 1(1), 2–30.

Gaddimath, S., Payamalle, S., Channabasavana Hundi Puttaningaiah, K. P., Hur, J. 2024. Recent advances in pH and redox responsive polymer nanocomposites for cancer therapy. *J. Compos. Sci.*, 8(1), 28.

Gafiullina, A., Ladewig, B. P., Zhang, J. 2022. Photoresponsive polymer and polymer composite membranes for gas separation. *ACS Appl. Polym. Mater.*, 5(1), 1–30.

Ge, H., Ma, Z. 2015. Microwave preparation of triethylenetetramine modified graphene oxide/chitosan composite for adsorption of Cr (VI). *Carbohydr. Polym.*, 131, 280–287.

Gouda, M. H., Gouveia, W., Elessawy, N. A., Šljukić, B., Nassr, A. B. A. A., Santos, D. M. F. 2020. Simple design of PVA-based blend doped with SO_4 (PO_4)-functionalised TiO_2 as an effective membrane for direct borohydride fuel cells. *Int. J. Hydrog. Energy*, 45(30), 15226–15238.

Harun-Ur-Rashid, M., Imran, A. B., Susan, M. A. B. H. 2023. Green polymer nanocmposites in automotive and packaging industries. *Curr. Pharm. Biotechnol.*, 24(1), 145–163.

He, J., Liang, Y., Shi, M., Guo, B. 2020. Anti-oxidant electroactive and antibacterial nanofibrous wound dressings based on poly (ε-caprolactone)/quaternized chitosan-graft-polyaniline for full-thickness skin wound healing. *Chem. Eng. J.*, 385, 123464.

Hosseini, M. G., Shahryari, E., Yardani Sefidi, P. 2021. Polyaniline grafted chitosan/GO-CNT/Fe_3O_4 nanocomposite as a superior electrode material for supercapacitor application. *J. Appl. Polym. Sci.*, 138(38), 50976.

Hsu, L., Weder, C., Rowan, S. J. 2011. Stimuli-responsive, mechanically-adaptive polymer nanocomposites. *J. Mater. Chem.*, 21(9), 2812–2822.

Irfan, M., Polonskyi, O., Hinz, A., Mollea, C., Bosco, F., Strunskus, T., Balagna, C., Perero, S., Faupel, F., Ferraris, M. 2019. Antibacterial, highly hydrophobic and semi transparent Ag/plasma polymer nanocomposite coating on cotton fabric obtained by plasma based co-deposition. *Cellulose*, 26, 8877–8894.

Islam, H. B. M. Z., Susan, M. A. B. H., Imran, A. B. 2020. Effects of plasticizers and clays on the physical, chemical, mechanical, thermal, and morphological properties of potato starch-based nanocomposite films. *ACS Omega*, 5(28), 17543–17552.

Jouault, N., Zhao, D., Kumar, S. K. 2014. Role of casting solvent on nanoparticle dispersion in polymer nanocomposites. *Macromolecules*, 47(15), 5246–5255.

Junior, J. P. C., Soares, I. L., Luetkmeyer, L., Tavares, M. I. B. 2014. Preparation of high-impact polystyrene nanocomposites with organoclay by melt intercalation and characterization by low-field nuclear magnetic resonance. *Chem. Eng. Process. Process Intensif.*, 77, 66–76.

Katowah, D. F., Alqarni, S., Mohammed, G. I., Al Sheheri, S. Z., Alam, M. M., Ismail, S. H., Asiri, A. M., Hussein, M. A., Rahman, M. M. 2020. Selective Hg^{2+} sensor performance based various carbon-nanofillers into CuO-PMMA nanocomposites. *Polym. Adv. Technol.*, 31(9), 1946–1962.

Kausar, A. 2016. Polymer/graphene nanocomposite: Preparation to application. *Am. J. Polym. Sci. Eng.*, 4(1), 111–122.

Khan, M. U. A., Razaq, S. I. A., Mehboob, H., Rehman, S., Al-Arjan, W. S., Amin, R. 2021. Antibacterial and hemocompatible pH-responsive hydrogel for skin wound healing application: In vitro drug release. *Polymers*, 13(21), 3703.

Khdary, N. H., Abdelsalam, M. E. 2020. Polymer-silica nanocomposite membranes for CO_2 capturing. *Arab. J. Chem.*, 13(1), 557–567.

Kim, J., Kim, T. H., Lee, J. H., Park, Y. A., Kang, Y. J., Ji, H. G. 2021. Porous nanocomposite of layered double hydroxide nanosheet and chitosan biopolymer for cosmetic carrier application. *Appl. Clay Sci.*, 205, 106067.

Kim, J., Lee, D., Lee, Y., Chen, W., Lee, S. 2019. Nanocellulose for energy storage systems: Beyond the limits of synthetic materials. *Adv. Mater.*, 31(20), 1804826.

Koolivand, H., Sharif, A., Kashani, M. R., Karimi, M., Salooki, M. K., Semsarzadeh, M. A. 2014. Functionalized graphene oxide/polyimide nanocomposites as highly CO_2-selective membranes. *J. Polym. Res.*, 21, 1–12.

Kumawat, M., Thapliyal, D., Verros, G. D., Arya, R. K., Barman, S., Halder, G., Shandilya, P. 2022. PANI-based hydrogen sulfide gas sensors. *Coatings*, 12(2), 186.

Lau, K., Gu, C., Hui, D. 2006. A critical review on nanotube and nanotube/nanoclay related polymer composite materials. *Compos. Part B Eng.*, 37(6), 425–436.

Lavrador, P., Esteves, M. R., Gaspar, V. M., Mano, J. F. 2021. Stimuli-responsive nanocomposite hydrogels for biomedical applications. *Adv. Funct. Mater.*, 31(8), 2005941.

Liu, J., Tao, L., Yang, W., Li, D., Boyer, C., Wuhrer, R., Braet, F., Davis, T. P. 2010. Synthesis, characterization, and multilayer assembly of pH sensitive graphene-polymer nanocomposites. *Langmuir*, 26(12), 10068–10075.

Loguercio, L. F., Thesing, A., Demingos, P., de Albuquerque, C. D. L., Rodrigues, R. S. B., Brolo, A. G., Santos, J. F. L. 2021. Efficient acetylcholinesterase immobilization for improved electrochemical performance in polypyrrole nanocomposite-based biosensors for carbaryl pesticide. *Sens. Actuators B: Chem.*, 339, 129875.

Lu, J., Wang, K., Lei, W., Mao, Y., Di, D., Zhao, Q., Wang, S. 2021. Polydopamine-carbon dots functionalized hollow carbon nanoplatform for fluorescence-imaging and photothermal-enhanced thermochemotherapy. *Mater. Sci. Eng. C*, 122, 111908.

Lu, Q., Jang, H. S., Han, W. J., Lee, J. H., Choi, H. J. 2019. Stimuli-responsive graphene oxide-polymer nanocomposites. *Macromol. Res.*, 27, 1061–1070.

Lyu, W., Wu, J., Zhang, W., Liu, Y., Yu, M., Zhao, Y., Feng, J., Yan, W. 2019. Easy separated 3D hierarchical coral-like magnetic polyaniline adsorbent with enhanced performance in adsorption and reduction of Cr (VI) and immobilization of Cr (III). *Chem. Eng. J.*, 363, 107–119.

Maiti, M., Bhattacharya, M., Bhowmick, A. K. 2008. Elastomer nanocomposites. *Rubber Chem. Technol.*, 81(3), 384–469.

Manzoor, Q., Farrukh, M. A., Qamar, M. T., Sajid, A., Aldossari, S. A., Manikandan, A., Iqbal, M. 2024a. Polymer-assisted synthesis of ternary magnetic graphene oxide nanocomposite for the adsorptive removal of Cr (VI) and Pb (II) ions. *Heliyon*, 10(15), 35204.

Manzoor, W., Munir, R., Noreen, S. 2024b. Metal oxide-doped polymer-based nanocomposite hydrogels: Batch adsorption of reactive red dye from aqueous solutions. *Iran. Polym. J.*, 33(3), 333–354.

Ma, P. C., Siddiqui, N. A., Marom, G., Kim, J. K. 2010. Dispersion and functionalization of carbon nanotubes for polymer-based nanocomposites: A review. *Compos. Part A Appl. Sci. Manuf.*, 41(10), 1345–1367.

Maruthupandy, M., Muneeswaran, T., Anand, M., Quero, F. 2020. Highly efficient multifunctional graphene/chitosan/magnetite nanocomposites for photocatalytic degradation of important dye molecules. *Int. J. Biol. Macromol.*, 153, 736–746.

Mashkoor, F., Shoeb, M., Jeong, C. 2023. Alginate modified magnetic polypyrrole nanocomposite for the adsorptive removal of heavy metal. *Polymers*, 15(21), 4285.

Murugan, A., Siva, V., Shameem, A., Al-Sehemi, A. G. 2024. Biodegradable polymers-wrapped $CoMoO_4$ hybrid composites: A promising flexible solid-state supercapacitor electrode. *J. Mater. Sci. Mater. Electron.*, 35(16), 1041.

Nadeem, R., Manzoor, Q., Iqbal, M., Nisar, J. 2016. Biosorption of Pb (II) onto immobilized and native *Mangifera indica* waste biomass. *J. Ind. Eng. Chem.*, 35, 185–194.

Orellana, J., Moreno-Villoslada, I., Bose, R. K., Picchioni, F., Flores, M. E., Araya-Hermosilla, R. 2021. Self-healing polymer nanocomposite materials by Joule effect. *Polymers*, 13(4), 649.

Ramadan, M. A., Sharawy, S., Elbisi, M. K., Ghosal, K. 2020. Eco-friendly packaging composite fabrics based on *in situ* synthesized silver nanoparticles (AgNPs) and treatment with chitosan and/or date seed extract. *Nano-Struct. Nano-Objects*, 22, 100425.

Ray, S. S., Okamoto, M. 2003. Polymer/layered silicate nanocomposites: A review from preparation to processing. *Prog. Polym. Sci.*, 28(11), 1539–1641.

Rezk, A. I., Bhattarai, D. P., Park, J., Park, C. H., Kim, C. S. 2020. Polyaniline-coated titanium oxide nanoparticles and simvastatin-loaded poly (ε-caprolactone) composite nanofibers scaffold for bone tissue regeneration application. *Colloids Surf. B: Biointerfaces*, 192, 111007.

Roy, H. S., Islam, M. M., Mollah, M. Y. A., Susan, M. A. B. H. 2018. Polyaniline-NiO nanocomposites as dielectric materials. *Mater. Today: Proc.*, 5(7), 15267–15276.

Roy, H. S., Mollah, M. Y. A., Islam, M. M., Susan, M. A. B. H. 2018. Poly (vinyl alcohol) MnO_2 nanocomposite films as UV-shielding materials. *Polym. Bull.*, 75(12), 5629–5643.

Sandler, J., Kirk, J. E., Kinloch, I. A., Shaffer, M. S. P., Windle, A. H. 2003. Ultra-low electrical percolation threshold in carbon-nanotube-epoxy composites. *Polymer*, 44(19), 5893–5899.

Santos, R. M., Vilaverde, C., Cunha, E., Paiva, M. C., Covas, J. A. 2016. Probing dispersion and re-agglomeration phenomena upon melt-mixing of polymer-functionalized graphite nanoplates. *Soft Matter*, 12(1), 77–86.

Shang, L., Huang, Z., Pu, X., Yin, G., Chen, X. 2019. Preparation of graphene oxide-doped polypyrrole composite films with stable conductivity and their effect on the elongation and alignment of neurite. *ACS Biomater. Sci. Eng.*, 5(3), 1268–1278.

Shrestha, B. K., Shrestha, S., Baral, E. R., Lee, J. Y., Kim, B.-S., Park, C. H., Kim, C. S. 2019. Π-conjugated polyaniline-assisted flexible titania nanotubes with controlled surface morphology as regenerative medicine in nerve cell growth. *Chem. Eng. J.*, 360, 701–713.

Shu, G., Xu, D., Xie, S., Chang, L. J., Liu, X., Yang, J., Li, Y., Wang, X. 2023. The antioxidant, antibacterial, and infected wound Healing effects of ZnO quantum dots-chitosan biocomposite. *Appl. Surf. Sci.*, 611, 155727.

Singh, K., Jaiswal, R., Kumar, R., Singh, S., Agarwal, K. 2023. Polymer-based nanocomposites as defence material. *Bull. Mater. Sci.*, 46(2), 79.

Soto, G. D., Meiorin, C., Actis, D. G., Zélis, P. M., Londoño, O. M., Muraca, D., Mosiewicki, M. A., Marcovich, N. E. 2018. Magnetic nanocomposites based on shape memory polyurethanes. *Eur. Polym. J.*, 109, 8–15.

Stenzenberger, H. D. 1993. Recent developments of thermosetting polymers for advanced composites. *Compos. Struct.*, 24(3), 219–231.

Sun, W., Zhang, W., Li, H., Su, Q., Zhang, P., Chen, L. 2020. Insight into the synergistic effect on adsorption for Cr (VI) by a polypyrrole-based composite. *RSC Adv.*, 10(15), 8790–8799.

Taha, T. A. M., Alanazi, S. S., El-Nasser, K. S., Alshammari, A. H., Ismael, A. 2024. Structure-property relationships in PVDF/$SrTiO_3$/CNT nanocomposites for optoelectronic and solar cell applications. *Polymers*, 16(6), 736.

Tanahashi, M. 2010. Development of fabrication methods of filler/polymer nanocomposites: With focus on simple melt-compounding-based approach without surface modification of nanofillers. *Materials*, 3(3), 1593–1619.

Thakur, Y. S., Acharya, A. D., Sharma, S., Bisoyi, S. 2024. Enhanced electrochemical performance of *in situ* polymerized V_2O_5-PANI nanocomposites and its practical application confirmation by assembling ionic liquid as well as solid state-based supercapacitor device. *Results Chem.*, 7, 101259.

Thinh, D. B., Tien, N. T., Dat, N. M., Phong, H. H. T., Giang, N. T. H., Oanh, D. T. Y., Nam, H. M., Phong, M. T., Hieu, N. H. 2020. Improved photodegradation of p-nitrophenol from water media using ternary $MgFe_2O_4$-doped TiO_2/reduced graphene oxide. *Synth. Met.*, 270, 116583.

Van Berkum, S., Dee, J. T., Philipse, A. P., Erné, B. H. 2013. Frequency-dependent magnetic susceptibility of magnetite and cobalt ferrite nanoparticles embedded in PAA hydrogel. *Int. J. Mol. Sci.*, 14(5), 10162–10177.

Vera, M., Mella, C., Urbano, B. F. 2020. Smart polymer nanocomposites: Recent advances and perspectives. *J. Chil. Chem. Soc.*, 65(4), 4973–4981.

Wang, L., Yang, Y., Han, W., Ding, H. 2024. Novel design and development of *Centella asiatica* extract-loaded poloxamer/ZnO nanocomposite wound closure material to improve anti-bacterial action and enhanced wound healing efficacy in diabetic foot ulcer. *Regen. Ther.*, 27, 92–103.

Wang, N., Chen, J., Wang, J., Feng, J., Yan, W. 2019. Removal of methylene blue by polyaniline/TiO_2 hydrate: Adsorption kinetic, isotherm and mechanism studies. *Powder Technol.*, 347, 93–102.

Wang, S., Guan, S., Zhu, Z., Li, W., Liu, T., Ma, X. 2017. Hyaluronic acid doped-poly (3,4-ethylenedioxythiophene)/chitosan/gelatin (PEDOT-HA/Cs/Gel) porous conductive scaffold for nerve regeneration. *Mater. Sci. Eng. C*, 71, 308–316.

Wang, X., Li, Q., Xie, J., Jin, Z., Wang, J., Li, Y., Jiang, K., Fan, S. 2009. Fabrication of ultralong and electrically uniform single-walled carbon nanotubes on clean substrates. *Nano Lett.*, 9(9), 3137–3141.

Won, H. J., Ryplida, B., Kim, S. G., Lee, G., Ryu, J. H., Park, S. Y. 2020. Diselenide-bridged carbon-dot-mediated self-healing, conductive, and adhesive wireless hydrogel sensors for label-free breast cancer detection. *ACS Nano*, 14(7), 8409–8420.

Yadav, S., Asthana, A., Chakraborty, R., Jain, B., Singh, A. K., Carabineiro, S. A. C., Susan, M. A. B. H. 2020. Cationic dye removal using novel magnetic/activated charcoal/ β-cyclodextrin/alginate polymer nanocomposite. *Nanomaterials*, 10(1), 170.

Yadid, M., Feiner, R., Dvir, T. 2019. Gold nanoparticle-integrated scaffolds for tissue engineering and regenerative medicine. *Nano Lett.*, 19(4), 2198–2206.

Yang, C., Liu, Z., Chen, C., Shi, K., Zhang, L., Ju, X. J., Wang, W., Xie, R., Chu, L. Y. 2017. Reduced graphene oxide-containing smart hydrogels with excellent electro-response and mechanical properties for soft actuators. *ACS Appl. Mater. Interfaces*, 9(18), 15758–15767.

Yue, J., He, L., Tang, Y., Yang, L., Wu, B., Ni, J. 2019. Facile design and development of photoluminescent graphene quantum dots grafted dextran/glycol-polymeric hydrogel for thermoresponsive triggered delivery of buprenorphine on pain management in tissue implantation. *J. Photochem. Photobiol. B Biol.*, 197, 111530.

Yunphuttha, C., Midpanon, S., Marr, D. W. M., Viravathana, P. 2024. Polyvinyl alcohol/nanocellulose nanocomposites from oil palm empty fruit bunch as anion exchange membranes for direct alcohol-hydrogen peroxide fuel cells. *Cellulose*, 31(3), 1569–1601.

Zahid, M., Heredia-Guerrero, J. A., Athanassiou, A., Bayer, I. S. 2017. Robust water repellent treatment for woven cotton fabrics with eco-friendly polymers. *Chem. Eng. J.*, 319, 321–332.

Zhang, Y., Ni, S., Wang, X., Zhang, W., Lagerquist, L., Qin, M., Willför, S., Xu, C., Fatehi, P. 2019. Ultrafast adsorption of heavy metal ions onto functionalized lignin-based hybrid magnetic nanoparticles. *Chem. Eng. J.*, 372, 82–91.

Zhao, W., Zhang, F., Leng, J., Liu, Y. 2019. Personalized 4D printing of bioinspired tracheal scaffold concept based on magnetic stimulated shape memory composites. *Compos. Sci. Technol.*, 184, 107866.

Index

A

Adsorbent, 116, 127
Adsorption, 116, 117, 122, 123, 127–131, 138
Agglomeration, 17, 38, 71, 152, 173
Alginate beads, 132, 145
Anion exchanger, 127, 129, 132
Anti-fouling, 136, 137
Antioxidant, 158, 159, 161, 169, 200, 204
Automobile, 4, 14, 15, 17, 31, 32, 68, 70

B

Ball milling, 8, 37, 39, 40, 173
Bandgap, 16, 34, 80, 84–87, 90
Biocompatibility, 11, 12, 14, 16, 17, 23, 47, 70, 147, 153, 168, 169, 177, 178, 198, 200, 204, 207, 209
Biodegradable, 16, 149, 150, 152–154, 176, 180, 188, 190
Birefringence, 28
Bulletproof, 189

C

Carbon nanostructure, 8, 16, 29, 33
Carbox methyl cellulose, 28, 130
Chemical functionalities, 147
Chemiresistive, 194
Chitosan, 17, 44, 120, 130, 133–135, 147, 153, 155, 156, 190, 197, 200, 204, 207
Classifications, 4, 5, 23, 120, 148, 189, 190
Clay, 2–4, 6, 12, 14, 15, 33, 36–38, 40–42, 45, 46, 60, 64, 68, 69, 71, 103, 105, 109, 123, 124, 147–150, 168, 169, 173, 175, 191–193, 197
Colloidal, 8, 44, 150
Commercialization, 17, 72, 177, 179, 181
Composite, 52, 119
Compounding, 7, 24, 36, 40, 71, 125, 126, 193
Computational, 60
Conducting polymers, 15, 25, 28, 29, 32, 125
Conductive filler, 98, 100
Copolymerization, 82
Creep, 60, 61, 64, 134

D

Dielectric capacitors, 28, 30, 85–89, 93, 99, 104, 107, 108, 201, 203
Dispersed phase, 2, 23, 118
Dope solution, 45, 46
Dye, 89, 116, 122, 127, 133, 134, 136, 138, 154–157, 170, 193, 196, 197

E

Economic, 42, 138, 147, 166, 167, 169, 174, 180
Electrical insulation, 29, 30
Electrically conductive polymers, 32
Electrohydrodynamic, 44
Electronics and electrical industries, 176
Electrospinning, 7, 8, 37, 44–46, 137, 153, 172
Environmental, 3, 22, 69, 72, 92, 100–102, 116, 127, 132, 138, 146–152, 154, 155, 161, 169, 170, 175, 179–181, 193, 194, 206–209
Exfoliated, 192, 193

F

Fertilizers, 16, 51
Flexural, 10, 29, 58–60
Fluorescence, 89–91, 201
Functional, 116
Functionalization, 5, 53, 68, 71, 72, 79, 98, 120, 121, 123, 128, 168, 169, 171, 177

G

Gas, 148
Green, 150, 152

H

Hardness, 11, 32, 54, 63, 64, 68, 71, 168, 169
Healing, 180–182, 200, 201, 207
Herbicides, 134, 136, 138
Homogenisation, 173
Humidity, 33
Hydrogels, 37, 48, 71, 118, 170, 201, 205
Hydrophobicity, 122, 136, 137, 160, 203

I

Impact, 29, 61, 62
Impulse, 106
Injectable, 206
In situ polymerization, 123, 171, 172, 177, 192, 206, 207
Intercalated, 36, 37, 40, 42, 125, 126, 152, 192
Interface, 2, 3, 9, 17, 37, 79, 107, 118, 146, 188
Ion exchange, 38, 116, 123, 127, 138, 195, 202

L

Laser irradiation, 200
Layer-by-layer deposition, 48
Layered silicates, 38, 68, 169
LB techniques, 8
Life Cycle Assessment, 154

M

Macromolecular, 21
Magneto, 205
Manufacturing processes, 208
Market analysis, 171
Master batch, 42
Maxwell–Garnett theory, 88, 89
Maxwell model, 54, 55
Mechanical, 4, 9–12, 14, 22, 28, 29, 33, 46, 47, 51–53, 65–70, 72, 81, 82, 92, 99, 101, 147, 149, 150, 153, 168, 170, 173, 175, 178, 189, 190, 200, 201, 203, 205
Medical devices, 14, 16, 23, 47, 70, 149, 168–170, 177, 198
Melt grafting, 27
Membranes, 17, 30, 118, 121, 123, 136, 137, 169, 200, 202
Metamaterials, 86, 87
Microwave, 98, 124, 126, 153
Monomer, 5, 6, 15, 20, 22–24, 38, 40–43, 67, 123, 125, 126, 151, 152, 171, 172, 192, 193, 203
Montmorillonites, 37

N

Nanocarbon, 66, 123, 124, 130
Nanofillers, 30, 37, 38, 40–42, 46, 51–53, 57, 58, 60, 63–68, 70–72, 80, 98, 102–104, 119, 148–150, 154, 161, 168, 170–173, 189–194, 205, 206
Nanomaterials, 2, 6, 9, 10, 16, 48, 51, 53, 80, 82, 93, 102, 103, 119, 137, 147, 149, 151, 155, 157, 167–169, 175, 179, 208
Nano revolution, 188, 209
Nanotechnology, 2, 14, 16, 88, 97, 102, 146, 166, 167, 171, 174, 180, 188, 208, 209
Natural, 2
Neural, 198, 199
Nonlinear, 85

O

Optical characteristics, 27, 28, 78, 80–82, 84, 88, 91–93, 170
Optoelectronic, 80, 102, 104
Organo-modified clay, 40

P

Permittivity, 83, 86, 107
Personalized medicine, 177
Pesticides, 134, 136, 138, 194
Phosphate, 1, 122, 125, 132
Photocatalytic degradation, 133, 155, 197
Photoluminescence, 88, 90
Photonic, 85–87, 89, 93, 170
Photosensitive, 206
Physicochemical properties, 51, 208
Pollutants, 116, 118, 131, 133–138, 193, 194
Poly(acrylamide), 203
Polyhydroxyalkanoate, 149, 152
Polymer, 6, 7, 24, 28, 30, 32, 36, 37, 40–43, 55, 64, 66–68, 71, 72, 78, 79, 81–83, 88, 89, 91–93, 97, 98, 100, 108, 117–120, 122–126, 146–148, 150–152, 154, 166, 168, 170–172, 177, 190–193, 200, 202, 205, 206
Polymerization, 22–24
Power Cable, 105–107, 109

Q

Quantum, 79, 80, 84, 88, 91–93, 103, 167, 169, 191

R

Refractive index, 27, 28, 78, 80–89, 91–93, 203
Regeneration, 70, 129, 132, 133, 138, 195, 198–200
Regenerative therapies, 177
Regulations, 175, 179–182
Regulatory compliance, 167, 179
Reinforcement, 4, 64–68, 118, 119, 122, 147, 148, 168–170
Renewable energy storage, 157
Renewable feedstocks, 151, 152
Renewable resources, 151, 154, 158, 208

S

Scalability, 7, 14, 17, 66, 167, 179
Scale-up process, 72
Self-healing nanocomposite, 180, 181
Semiconductors, 84, 90, 103
Shear relaxation, 64
Shear strength, 62, 63
Shelf life, 69, 176, 178
Sol-gel process, 6, 8, 43, 44, 46, 48, 172
Solution based, 3d printing, 48
Solvent-free processes, 151
Specific surface area, 67, 119, 121, 122, 146
Spectrometers, 26
Spray pyrolysis, 150

Strategic initiatives, 174, 175
Stress–strain, 53–57, 59–61
Sunscreens, 168, 169
Supercapacitor, 80, 103, 104, 157, 158, 201, 202, 207
Surface area, 3, 13, 22, 36, 45, 51, 65–68, 80, 100, 102, 119, 121–123, 127, 136, 137, 146, 147, 153, 155, 168–170, 172, 190, 194, 195, 198, 204
Surface functionalization, 53, 68, 71, 72, 79, 98, 171
Sustainability, 2, 116, 120, 129, 148, 154, 167, 176, 178–181, 201, 203
Sustainable environment, 194

T

Template synthesis, 8, 40, 42, 125, 126, 170, 193
Tensile strength, 3, 29, 51, 52, 54, 55, 57, 58, 63, 65–71
Thermal conductivity, 13, 15, 100, 101
Thermal management, 100
Thermoplastics, 125, 148, 149, 174, 175, 190
Thermoset, 36, 148, 149, 158, 190
3D printing, 46–48, 173, 177, 180, 181
Toughness, 12, 14, 29, 51, 52, 54, 61–63, 65, 71, 101, 118, 170
Transparency, 13, 28, 78, 82, 91, 92, 203

U

Universal tensile testing machine, 56
UV protection, 3, 101, 168

W

Water management practices, 155
Water pollution, 116, 146
Water purification, 116–118, 122, 127, 154
Water-repellent properties, 161
Wound healing, 198, 200

X

X-ray diffractometer, 26
X-ray photoelectron spectroscopy, 9